DESIGN OF LOW-VOLTAGE BIPOLAR OPERATIONAL AMPLIFIERS

THE KLUWER INTERNATIONAL SERIES IN ENGINEERING AND COMPUTER SCIENCE

ANALOG CIRCUITS AND SIGNAL PROCESSING

Consulting Editor

Mohammed Ismail
Ohio State University

Related titles:

STATISTICAL MODELING FOR COMPUTER-AIDED DESIGN OF MOS VLSI CIRCUITS, Christopher Michael, Mohammed Ismail
ISBN: 0-7923-9299-X
SELECTIVE LINEAR-PHASE SWITCHED-CAPACITOR AND DIGITAL FILTERS, Hussein Baher
ISBN: 0-7923-9298-1
ANALOG CMOS FILTERS FOR VERY HIGH FREQUENCIES, Bram Nauta
ISBN: 0-7923-9272-8
ANALOG VLSI NEURAL NETWORKS, Yoshiyasu Takefuji
ISBN: 0-7923-9273-6
ANALOG VLSI IMPLEMENTATION OF NEURAL NETWORKS, Carver A. Mead, Mohammed Ismail
ISBN: 0-7923-9040-7
AN INTRODUCTION TO ANALOG VLSI DESIGN AUTOMATION, Mohammed Ismail, Jose Franca
ISBN: 0-7923-9071-7
INTRODUCTION TO THE DESIGN OF TRANSCONDUCTOR-CAPACITOR FILTERS, Jaime Kardontchik
ISBN: 0-7923-9195-0
VLSI DESIGN OF NEURAL NETWORKS, Ulrich Ramacher, Ulrich Ruckert
ISBN: 0-7923-9127-6
LOW-NOISE WIDE-BAND AMPLIFIERS IN BIPOLAR AND CMOS TECHNOLOGIES, Z. Y. Chang, Willy Sansen
ISBN: 0-7923-9096-2
ANALOG INTEGRATED CIRCUITS FOR COMMUNICATIONS: Principles, Simulation and Design, Donald O. Pederson, Kartikeya Mayaram
ISBN: 0-7923-9089-X
SYMBOLIC ANALYSIS FOR AUTOMATED DESIGN OF ANALOG INTEGRATED CIRCUITS, Georges Gielen, Willy Sansen
ISBN: 0-7923-9161-6
STEADY-STATE METHODS FOR SIMULATING ANALOG AND MICROWAVE CIRCUITS, Kenneth S. Kundert, Jacob White, Alberto Sangiovanni-Vincentelli
ISBN: 0-7923-9069-5
MIXED-MODE SIMULATION: Algorithms and Implementation, Reseve A. Saleh, A. Richard Newton
ISBN: 0-7923-9107-1

DESIGN OF LOW-VOLTAGE BIPOLAR OPERATIONAL AMPLIFIERS

by

M. Jeroen Fonderie
Philips Semiconductor/Signetics Company

Johan H. Huijsing
Delft University of Technology

KLUWER ACADEMIC PUBLISHERS
Boston/Dordrecht/London

Distributors for North America:
Kluwer Academic Publishers
101 Philip Drive
Assinippi Park
Norwell, Massachusetts 02061 USA

Distributors for all other countries:
Kluwer Academic Publishers Group
Distribution Centre
Post Office Box 322
3300 AH Dordrecht, THE NETHERLANDS

Library of Congress Cataloging-in-Publication Data

Fonderie, M. Jeroen, 1960-
Design of low-voltage bipolar operational amplifiers / by M. Jeroen Fonderie, Johan H. Huijsing.
p. cm. -- (The Kluwer international series in engineering and computer science ; SECS 218. Analog circuits and signal processing)
Includes bibliographical references and index.
ISBN 0-7923-9317-1 (alk. paper)
1. Operational amplifiers--Design and construction. 2. Bipolar transistors. I. Huijsing, Johan H., 1938- . II. Title.
III. Series: Kluwer international series in engineering and computer science ; SECS 218. IV. Series: Kluwer international series in engineering and computer science. Analog circuits and signal processing.
TK7871 . 58 . 06F66 1993
621 . 395--dc20 92-44955
CIP

Printed on acid-free paper.

Printed in the United States of America

CONTENTS

LIST OF NOTATIONS

symbol	quantity	unit
β	transistor current gain	-
$\omega_n^{(m1)}$	the n-th pole of the OpAmp with C_{m1} inserted	rad
$\omega_n^{(m2)}$	the n-th pole of the OpAmp with C_{m2}inserted	rad
$\omega_n^{(mp1)}$	the n-th pole of the MPD OpAmp with C_{m2}inserted	rad
$\omega_n^{(mp2)}$	the n-th pole of the MPD OpAmp with C_{m2}inserted	rad
$\omega_n^{(nc)}$	the n-th pole of the not-compensated OpAmp	rad
$\omega_n^{(u)}$	the n-th pole of the OpAmp with unity-gain feedback	rad
ω_{NPN}	transit frequency of n-p-n transistor	rad/s
ω_{PNP}	transit frequency of p-n-p transistor	rad/s
ω_T	transit frequency of transistor, general	rad/s
A	gain, general	-
$A\beta$	loop gain	-
A_0	low-frequency gain	-
A_i	current gain	-
C	capacitor	F
C_{be}	base-emitter capacity of transistor	F
C_{m1}	Miller capacitor, compensating output and intermediate stage	F
C_{m2}	Miller capacitor, compensating output, intermediate and input stage	F
f_{NPN}	transit frequency of n-p-n transistor	Hz
f_{PNP}	transit frequency of p-n-p transistor	Hz
f_T	transit frequency of transistor, general	Hz
g	transconductance, general	mho, Ω^{-1}
g_i	input-stage transconductance	mho, Ω^{-1}
g_m	intermediate-stage transconductance	mho, Ω^{-1}
g_o	output-stage transconductance	mho, Ω^{-1}
I	current	A
$\overline{i_{ni}^2}$	squared equivalent input noise current	A^2/Hz
I_{OS}	offset current	A

I_S	saturation current of transistor	A
k	Boltzmann's constant, 1.3805×10^{-23}	J/K
p_n	the n-th pole in the frequency response	rad
q	electron charge, 1.6×10^{-19}	C
Q	symbol for transistor	-
R	symbol for resistor	Ω
r_{be}	base-emitter resistance: $r_{be} = \beta r_e$	Ω
r_e	emitter resistance: $r_e = kT/qI$	Ω
r_i	input resistance of amplifier (stage)	Ω
r_o	output resistance of amplifier (stage)	Ω
s	complex frequency	rad
T	temperature	K
U_c	common-mode input voltage	V
U_d	differential-mode input voltage	V
U_i	small-signal input voltage	V
$\overline{u_{ni}^2}$	squared equivalent input noise voltage	V^2/Hz
U_o	small-signal output voltage	V
V_{OS}	offset voltage	V
V_T	thermal voltage, kT/q = 26 mV, at 27 °C	V
z_n	the n-th zero in the frequency response	rad

acronyms:

BPR	Bandwidth-to-Power Ratio
CE	Common Emitter
CM	Common Mode
CMRR	Common-Mode-Rejection Ratio
DC	Direct Current
HF	High Frequency
MPD	Multi-Path Driven
OpAmp	Operational Amplifier
PFCC	Parallel-Feedback-Current Compensated

PREFACE

This book deals with the design and implementation of low-voltage bipolar Operational Amplifiers (OpAmps). It is the result of the research that was started about ten years ago by Prof. Johan H. Huijsing. In 1987, he was joined by M. Jeroen Fonderie, who then began his four-year Ph.D. study. This search for better ways to implement low-voltage operational amplifiers has been carried out at the Delft University of Technology, the Netherlands, in cooperation with Philips Semiconductors, Signetics, Sunnyvale, CA. The material presented in this book will be of particular interest to professional designers of integrated amplifiers and to graduate students in this field. It is assumed that the reader has a basic knowledge of analog electronics and integrated circuit technology.

In Chapter 1, the question is answered why a low-voltage design is examined and why the bipolar technology is chosen. Further, the objectives of this work are defined and some design considerations are given. In Chapter 2, input stages are discussed. The input stage of an OpAmp should amplify the differential-mode input voltage, independent of the common-mode input voltage. This common-mode input voltage must be able to have any value that fits within the supply voltage range. Input stages that are able to realize this at supply voltages down to 1.8 V and down to 1 V are discussed. Chapter 3 deals with output stages. The output voltage of the OpAmp must be able to have any value within the supply voltage range. Besides this rail-to-rail output-voltage range, the output stage should also be able to supply the load that is connected to the OpAmp with an output current in the order of 10 mA. The current gain should therefore be as great as possible. The frequency response of the output stage should be very good, because the frequency behavior of the complete OpAmp is directly dependent on that of the output stage. In addition to the input and output stage, the OpAmp comprises several other circuit parts. The intermediate stage, between the input stage and the output stage boosts the overall gain of the OpAmp. The class-AB current control is required to efficiently control the currents in the output stage. The protection circuits prevent destruction of the OpAmp if it is driven beyond its specification limits. And finally the PTAT reference circuit is required to define all the currents that flow through the OpAmp. All these circuit parts are briefly discussed in Chapter 4. The OpAmp consists of several stages in series and each of these stages introduces at least one pole frequency in the transfer function of the OpAmp. If feedback is applied to the OpAmp, it oscillates if these poles are too close to each other, and, therefore, a frequency-compensation technique is used to split the poles apart. Both this Miller compensation and a new technique, the multi-path-driven Miller compensation, are thoroughly analyzed in Chapter 5. In Chapter 6, several realizations are evaluated and the measurements that have been performed to verify the theory are discussed. A Figure of merit, the bandwidth-to-power ratio, is defined in order to be able to compare the performance of the realized OpAmps. In the final Chapter, the major conclusions of this book are recapitulated and some references to further research are given.

M. Jeroen Fonderie
Johan H. Huijsing

DESIGN OF LOW-VOLTAGE BIPOLAR OPERATIONAL AMPLIFIERS

CHAPTER 1
INTRODUCTION

This work deals with the design of OpAmps, and, more specifically, with the design of bipolar OpAmps that are able to function at low supply voltages. An OpAmp is a universal amplifier that ideally amplifies its differential input voltage with an infinite gain to a single output. It has an infinite input resistance and a zero output resistance. Because of this infinite gain, the use of the OpAmp requires the application of external feedback. Fig. 1.1 shows the symbol of an OpAmp with a passive feedback circuit.

Although OpAmps were used primarily in analog computers, years before the invention of the integrated circuit, their application in system design has only become widespread over the last three decades, since monolithic-integration technology allowed the production of the OpAmps on a large scale and at a low

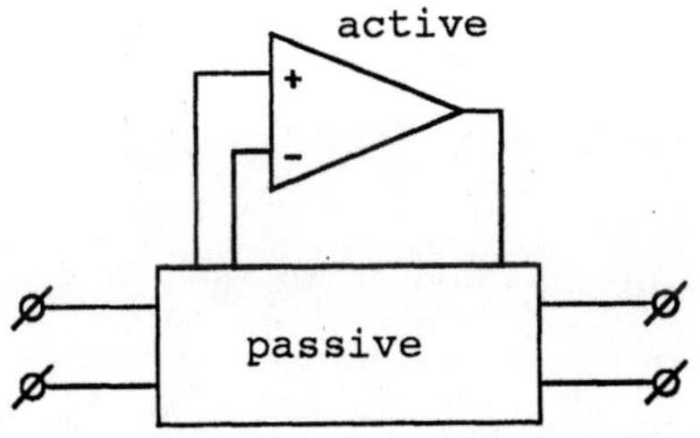

Fig. 1.1. *Operational Amplifier connected as non-inverting voltage amplifier.*

cost. Today we see that numerous OpAmps have been designed, and although it still is basically a general-purpose device, an OpAmp can be chosen that best suits the specific application. The OpAmp can be used as sensor-interface amplifier, as anti-aliasing filter in front of an analog-to-digital converter, as post-sampling filter behind a digital-to-analog converter and as power amplifier. Further, the OpAmp is a general-purpose device that can be used as filter, shaper and buffer in any analogue system design. There are OpAmps that are able to operate at very high supply voltages [1], and ones that are suited for use at very low supply voltages [2], [3], [4], [5]. OpAmps with an extremely low power consumption have been designed [6], [7] and also high-power OpAmps [8], [9]. There are OpAmps with a very high bandwidth [10], [11], [12] and OpAmps that are almost noise and offset free [13], [14]. These OpAmps are mainly realized in bipolar processes, but in the last decennium, high-performance CMOS OpAmps have also been reported [15], [16], [17], [18], [19], [20].

The first two Sections of this Introduction answer the question why a bipolar low-voltage design is chosen. The objectives of this work are defined in Section 1.3 and in Section 1.4 OpAmp design considerations are discussed. Section 1.5, finally, outlines the organization of this book.

1.1 WHY USE A LOW SUPPLY VOLTAGE?

Why use a low supply voltage? First of all, there is a tendency towards chip components of smaller dimensions. A component with these smaller dimensions is subject to breakdown at lower voltages and, therefore, this factor impels the use of low supply voltages. The need to keep the power consumption per chip area low necessitates the use of both low voltages and low currents.

Secondly, battery-powered portable equipment should be able to function at low supply voltages, preferably about 1 V, using just one single battery cell up to the end of its life span.

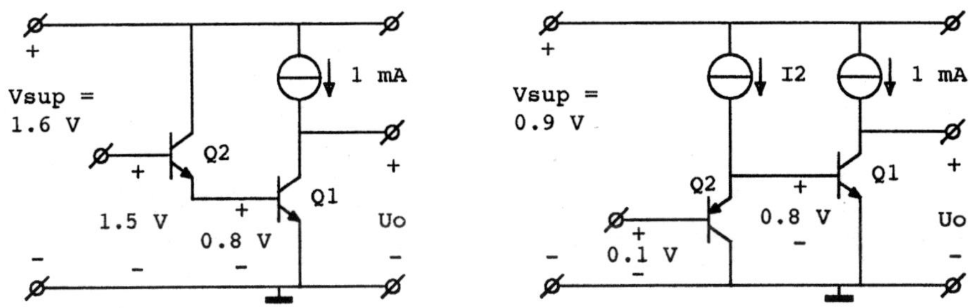

Fig. 1.2. Circuit with a minimum supply voltage of 1.5 V (left) and with a minimum supply voltage of 0.9 V (right).

A third reason can be found in the improvement of the general-purpose character of the OpAmp. If low supply voltages are allowed, applying it in system design is less critical; the circuit can, for instance, be made to operate from residual voltages within the system [21].

Although most analogue functions can be implemented even at a supply voltage as low as 1 V, the resulting circuits invariably use more circuit parts and more quiescent current. To illustrate this, Fig. 1.2 is shown. On the left a Darlington pair is depicted, as could, for instance, be found at the output of a linear circuit. At a collector current of 1 mA, the base-emitter voltage of Q_1 is in the order of 800 mV. The driver transistor Q_2 is biased at a much smaller current, so we will assume its base-emitter voltage to be 700 mV. The sum of both is then 1.5 V. At this voltage level, the preceding transistor still has to be able to drive Q_2, so at least 100 mV should be added to the 1.5 V to obtain the minimum supply voltage. Consequently, the Darlington configuration, and, in fact, any configuration with two stacked transistors, requires a supply voltage of minimally 1.6 V.

On the right of Fig. 1.2, a folded Darlington is shown. The driver Q_2 is now a p-n-p transistor, and, therefore, the input-voltage level of the circuit is shifted down to 0.1 V. Because there must to be a minimum voltage of 100 mV over current source I_2, the minimum supply voltage here is 0.9 V. Although both circuits of Fig. 1.2 basically perform the same function, the amount of current that can be sunk by Q_1 of the folded Darlington is restricted by current source I_2. Moreover, the voltage at the base of Q_2 should be at least 0.1 V, again, because the preceding transistor has to be able to drive Q_2. This requires Q_2 to have a low base-emitter voltage and, consequently, a large emitter area. This example demonstrates the several constraints of low-voltage designs: the use of p-n-p transistors is inevitable, high current levels are more difficult to obtain and the circuit occupies a larger chip area and consumes more quiescent current.

1.2 WHY USE BIPOLAR TECHNOLOGY?

The question might arise as to whether the OpAmp could better be composed of MOS transistors. After all, MOS transistors, potentially, can be made with a very low threshold voltage. A disadvantage of MOS transistors is, however, that their transconductance is much smaller than that of bipolar transistors. An increase in the drain current of a factor of 10 can easily cause a doubling of the gate-source voltage. The same increase in the collector current of a bipolar transistor causes only an enlargement of 60 mV of its base-emitter voltage. As is shown in Chapter 3, the bandwidth of an OpAmp is directly coupled to the transconductance of the output stage, and thus a bipolar OpAmp, potentially, has a higher bandwidth than an OpAmp that consists of MOS transistors. A large transconductance of the input stage is advantageous to obtain low values of input offset and noise voltage.

Furthermore, the minimum drain-source voltage of MOS transistors that still corresponds to a high drain-source impedance is usually much larger than the corresponding collector-emitter voltage of a bipolar transistor. As illustrated in the previous Section, this quantity, too, determines the minimum supply voltage. At first glance, the employment of bipolar technology therefore seems to be more advantageous. Nevertheless, MOS transistors can be employed in low-voltage design [22]. This work, however, is restricted to the use of bipolar transistors.

1.3 OBJECTIVES OF THE PRESENT WORK

The first 1-V OpAmp was designed in 1978 [2], so the objective of the present work on low-voltage OpAmps extended to more than just the realization of any OpAmp able to operate at a low supply voltage. The design preferably had to include the following features:

- a rail-to-rail input-voltage range
- a rail-to-rail output voltage range
- an overall voltage gain of more than 100 dB
- an output-current drive in the order of ±10 mA
- a bandwidth in the Mega-Hertz range
- a total current consumption of less than 1 mA

In order to achieve the stated aims, an inventory of ways to implement the input stage, the intermediate stage, the output stage and the frequency-compensation method first has to be made. This intends to bring to light the expense at which each of the above-mentioned features can be realized. After the completion of the inventory realization of a straight-forward OpAmp design should be possible, in which the pros and cons of each of the requirements can be considered, depending on the proposed application of the OpAmp.

1.4 DESIGN CONSIDERATIONS

Fig. 1.3 shows the equivalent circuit of an OpAmp, with the addition of circuit elements that characterize the deviations from its ideal behavior [23]. All these faults deteriorate the accurate performance of the OpAmp when it is used in conjunction with other circuit parts. When designing an OpAmp the necessity of keeping the imperfections as small as possible should be kept firmly in mind. Each

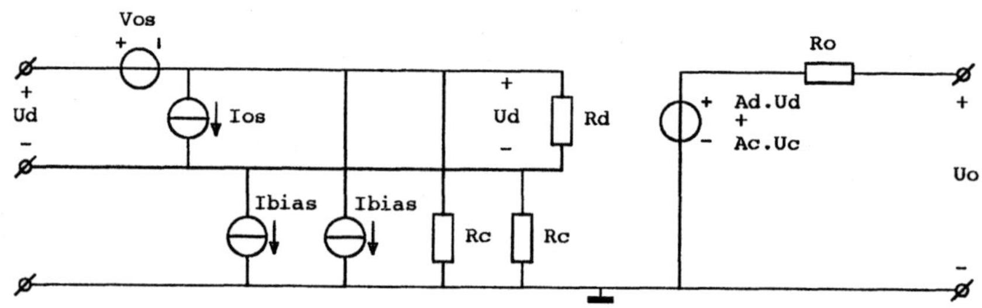

Fig. 1.3. Equivalent circuit of an operational amplifier with circuit elements to symbolize non-ideal behavior.

of the non-idealities of the OpAmp and its origin will now be discussed.

The OpAmp consists of several stages of bipolar or MOS transistors in series, and the overall gain can therefore be very high. Because of thermal feedback from the output transistors to the input, the gain is limited to a value of about 120 dB [24]. In Fig. 1.3, the driven voltage source with the finite gain A_d symbolizes the amplifying character of the OpAmp.

The input of an OpAmp usually consists of a differential pair of bipolar transistors. The base current of these transistors returns as the input bias current of the OpAmp and is symbolized by the two equal current sources I_{BIAS} in Fig. 1.3. This input bias current causes an additive error in the output voltage of the OpAmp.

If the current gain β of the transistors at the input of the OpAmp are not equal, the difference in input currents is the input offset current I_{OS}. The current source I_{OS} in Fig. 1.3 can also be thought to comprise the equivalent noise input current of the OpAmp. the current I_{OS} also induces an additive error.

Any mismatch in the transistors at the input of the OpAmp causes an offset voltage difference V_{OS} between the inputs. The equivalent input noise voltage can also thought to be included in the source V_{OS}. The input offset voltage causes an additive error.

The common-mode input voltage influences the result at the output of the OpAmp. This crosstalk of the common-mode input voltage on the differential-mode input voltage is called the common-mode rejection ratio (*CMRR*) and is equal to the ratio of the differential-mode gain to the common-mode gain. It causes a multiplicative error in the output voltage of the OpAmp. The common-mode gain is indicated in Fig. 1.3 by A_c.

The input resistance of the OpAmp is determined by the active devices at the input of the OpAmp. Particularly when bipolar transistors are used, the input resistance is certainly not infinite. When feedback is applied in series with the input, and the OpAmp has sufficient gain, the resulting input resistance is, however, very high. The input resistance of the OpAmp, represented by R_d, gives a multiplicative error.

Although the differential input is assumed to be isolated from ground, this is not the case in practice. Normally, the output resistance of the tail-current source that supplies the input pair limits this isolation. This common-mode input resistance is indicated by R_c and it causes a multiplicative error.

Sometimes, the output of the OpAmp is the emitter or source of the output transistor or transistors. The output resistance then may be low, but it is not zero. The collector or drain of the output transistor(s) may also be the output of the OpAmp, and then the output resistance is even rather high. Again, however, if the gain of the OpAmp is high enough, the feedback that is externally applied usually provides the OpAmp with a low output resistance. This output resistance, indicated by R_o, too, causes a multiplicative error.

Ideally, the voltage gain of the OpAmp is independent of the frequency of the input signal. Because of the limited transit frequencies of the active components of which the OpAmp consists, the OpAmp must be internally compensated if it is to be stable under all feedback conditions. Practical gain therefore decreases with the frequency, and eventually becomes one at the unity-gain bandwidth. Other characteristics of the OpAmp like the input and output impedance are also dependent on the frequency. This is not indicated in Fig. 1.3.

If large input signals are supplied to the OpAmp, the input stage does not operate linearly any more. In particular, a step input voltage results in a ramp output voltage. This process of slewing is not symbolized in Fig. 1.3.

The OpAmp is supplied by one or two main voltage sources of limited value. Consequently, the voltages at the input and output can at best have maximal values that are equal to these supply voltages. To protect the OpAmp from destroying itself, the output current that can be pushed or pulled from the load is limited. This too is not indicated in Fig. 1.3.

1.5 OUTLINE OF THIS BOOK

The contents of this book are divided into seven chapters. After the introduction in this Chapter, three Chapters are dedicated to the discussion of the sections that compose the OpAmp.

Chapter 2 deals with the design of input stages for low-voltage OpAmps. The purpose of this design emanates from the objective of having a common-mode input-voltage range that reaches from one supply rail to the other, and designs that have this feature both at a 2-V and at a 1-V supply are discussed.

In Chapter 3, possible output-stage configurations are analyzed. This discussion is restricted to output stages that have an output-voltage range that also reaches from rail to rail. Further, the output stage should be able to supply a sufficiently large output current to the load that is externally connected to the OpAmp, and, finally, the discussion in this Chapter focuses on the frequency response of the output stage.

The circuit parts that remain to complete the design of the OpAmp are discussed in Chapter 4. These circuit parts are the intermediate stage, inserted between the input and output stage to boost the overall gain of the OpAmp, some implementations of the class-AB current-control circuit, circuitry to protect the output transistors from heavy saturation, and from excessive power consumption, and, finally, the PTAT reference-current generator.

In Chapter 5, a detailed analysis is given of the frequency-compensation techniques that can be used to stabilize the OpAmp. From this theory, design criteria to successfully implement the compensation method are derived.

Chapter 6 describes the experimental OpAmp designs that have been realized, and the measurements that have been performed on these designs.

Finally, in Chapter 7, the conclusions of the preceding Chapters are summarized, and some suggestions for further research are given.

1.6 REFERENCES

[1] R. W. Russel, J. E. Solomon, "A High-Voltage Monolithic Operational Amplifier", *IEEE J. Solid-State Circuits*, vol. SC-6, pp. 352-357, Dec. 1971.

[2] R. J. Widlar, "Low-Voltage Techniques", *IEEE J. Solid-State Circuits*,, SC-13, pp. 838-846, Dec. 1978.

[3] S. D. S. Malhi, C. A. T. Salama and W. R. Donnison, "A Low-Voltage Micropower JFET/Bipolar Operational Amplifier", *IEEE J. Solid-State Circuits*, SC-16, pp. 669-676, Dec. 1981.

[4] O. H. Schade, Jr. and E. J. Kramer, "A Low-Voltage BiMos Op Amp", *IEEE J. Solid-State Circuits*, SC-16, pp. 661-668, Dec. 1981.

[5] J. H. Huijsing and D. Linebarger, "Low-Voltage Operational Amplifier with Rail-to-Rail Input and Output Ranges", *IEEE J. Solid-State Circuits*, SC-20, pp. 1144-1150, Dec. 1985.

[6] O. H. Schade, Jr., "BiMos Micropower IC's", *IEEE J. Solid-State Circuits*, SC-13, pp. 791-798, Dec. 1978.

[7] "OP-90 Precision Low-Voltage Micropower Operational Amplifier", *Precision Monolithics Inc. Technical Specification Sheet*, pp. 1-11, May 1987.

[8] P. R. Gray, "A 15-W Monolithic Power Operational Amplifier", *IEEE J. Solid-State Circuits*, vol. SC-7, pp. 474-480, Dec. 1972.

[9] R. J. Widlar and M. Yamatake, "A Monolithic Power Op Amp", *IEEE J. Solid-State Circuits*, SC-23, pp. 527-535, April 1988.

[10] T. J. van Kessel, "An Integrated Operational Amplifier with Novel HF Behavior", *IEEE J. Solid-State Circuits*, vol. SC-3, pp. 348-352, Dec. 1968.

[11] R. J. van de Plassche, "A Wide Band Operational Amplifier with a New Output Stage and Simple Frequency Compensation", *IEEE J. Solid-State Circuits*, vol. SC-6, pp. 347-352, Dec. 1971.

[12] J. H. Huijsing and F. Tol, "Monolithic Operational Amplifier Design with Improved HF Behavior", *IEEE J. Solid-State Circuits*, vol. SC-11, pp. 323-328, Dec. 1976.

[13] G. Erdi, "A Precision Trim Technique for Monolithic Analog Circuits" *IEEE J. Solid-State Circuits*, vol. SC-10, pp. 412-416, Dec. 1975.

[14] "Ultra-Low Offset Voltage Op Amp", *Precision Monolithics Inc. Technical Specifications Sheet*, pp. 1-8, Nov. 1977.

[15] P. R. Gray and R. G. Meyer, MOS Operational Amplifier Design- A Tutorial Overview", *IEEE J. Solid-State Circuits*, SC-17, pp. 969-962, Dec. 1982.

[16] R. Klinke, J. Hosticka and H. J. Pfleiderer, "A Very High Slew-Rate Dynamic CMOS Operational Amplifier", in *ESSCIRC Dig. Tech. Papers*, 1988, pp. 84-87.

[17] D. M. Monticelli, "A Quad CMOS Single-Supply Op Amp with Rail-to-Rail Output Swing", *IEEE J. Solid-State Circuits*, SC-21, pp. 1026-1034, Dec. 1986.

[18] M. Steyaert and W. Sansen, "A High-Dynamic-Range CMOS Op Amp with Low-Distortion Output Structure", *IEEE J. Solid-State Circuits*, SC-22, pp. 1204-1207, Dec. 1987.

[19] K. Bult and G. J. G. M. Geelen, "A Fast Settling CMOS Op Amp for SC Circuits with 90-dB DC Gain", *IEEE J. Solid-State Circuits*, SC-25, pp. 1379-1384, Dec. 1990.

[20] M. D. Pardoen and M. G. Degrauwe, "A Rail-to-Rail Input/Output CMOS Power Amplifier", *IEEE J. Solid-State Circuits*, SC-25, pp. 501-504, Dec. 1990.

[21] R. J. Widlar, R. Dobkin and M. Yamatake, "Low-voltage Op-Amp Breakthrough Expands Linear Design Horizons", *EDN*, pp. 91-99, Feb. 5, 1979.

[22] R. Hogervorst, R. J. Wiegerink, P. A. L. de Jong, J. Fonderie, R. F. Wassenaar and J. H. Huijsing, "CMOS Low-Voltage Operational Amplifiers with Constant g_m Rail-to-Rail Input Stage", in *Proceedings IEEE Int. Symp. on Circuits and Systems*, 1992, pp. 2876-2879.

[23] J. H. Huijsing, *Integrated Circuits for Accurate Linear Analogue Electrical Signal Processing*. Ph.D. Thesis, Delft University of Technology, Delft, the Netherlands 1981, pp. 32-33.

[24] J. E. Solomon, "The Monolithic Op Amp: A Tutorial Study", *IEEE J. Solid-State Circuits*, SC-3, pp. 314-332, Dec. 1974.

CHAPTER 2
INPUT STAGES

The task of the input stage of a low-voltage OpAmp is to amplify the differential input voltage. To obtain a reasonable ratio between the signal and additive interferences such as, for instance, noise, the common-mode (CM) input-voltage range should be as wide as possible, preferably reaching from the positive to the negative supply rail. Further, a rail-to-rail CM input-voltage range also enhances the general-purpose nature of the OpAmp.

In this Chapter, differential input stages are discussed with respect to their CM input-voltage range and the influence of interference signals such as offset, bias, drift, noise and CM crosstalk. In Section 2.1, a complementary input stage with a CM input-voltage range that extends beyond the supply rails is described. The next Section, 2.2, deals with a improvement of the complementary stage that besides a rail-to-rail CM input-voltage range, also incorporates a constant transconductance over this entire CM input-voltage range. In Section 2.3 an input stage is introduced that also features a constant transconductance at supply voltages as low as 1 V. Finally, in Section 2.4, an extension to the differential input stages is discussed that ensures the safe operation of common-mode input signals even beyond the supply rails.

2.1 COMPLEMENTARY INPUT STAGE

To obtain a rail-to-rail input-voltage range, the input stage can consist of two parts which have to be placed in parallel [1]. Fig. 2.1 shows this type of input stage. The input of n-p-n pair Q_1, Q_2 is able to reach the positive supply rail and

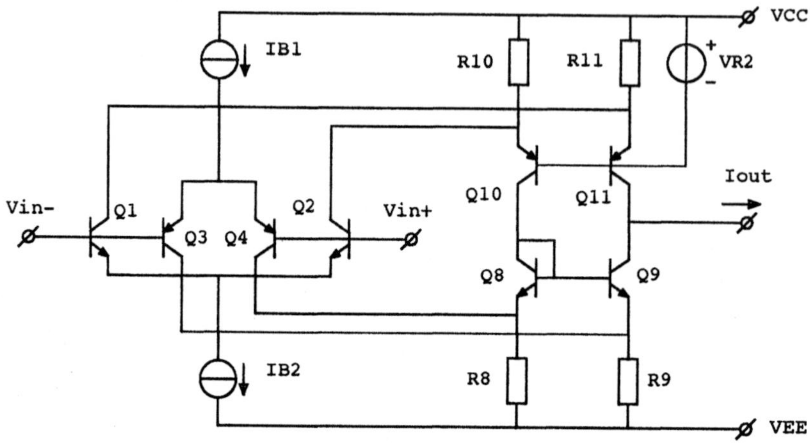

Fig. 2.1. Input stage with rail-to-rail common-mode input-voltage range.

the input of p-n-p pair Q_3, Q_4 can reach the negative supply rail. The voltage drop over resistors R_8-R_{11} must not be greater than 0.5 V, because otherwise the input transistors Q_1-Q_4 can become saturated when the CM-input voltage is near either one of the supply rails. The folded cascode transistors Q_8-Q_{10} sum the collector currents of the input transistors into one output current. Although in most OpAmps the input stage has a differential output, this does not essentially affect the functioning of the input stage shown in Fig. 2.1.

The saturation voltage over either one of the current sources I_{B1} or I_{B2}, together with the base-emitter voltages of either Q_1, Q_2 or Q_3, Q_4, determines the minimum supply voltage of this input stage at a value of approximately 0.1+0.7=0.8 V.

The common-mode ranges of the input stage as functions of the supply voltage, the so-called operational regions, are depicted in Fig. 2. This figure also shows that the minimum supply voltage at which the input stage has a rail-to-rail input-voltage swing is 1.6 V; below that either one of the input pairs is able to function near one of the supply rails, but in the middle of the CM range there is a gap where both are shut off because both I_{B1} and I_{B2} are saturated. If the supply voltage is higher than 1.6 V, the CM operation range of the input stage can be divided into three parts:

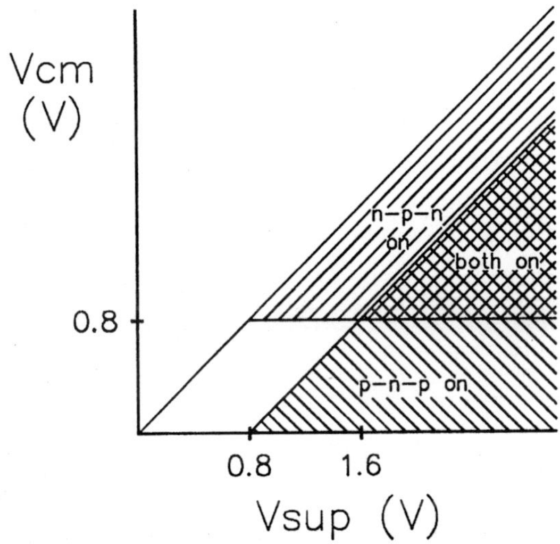

Fig. 2.2. The operational regions of the input stage shown in Fig. 2.1.

1. From the negative supply voltage V_{EE} to V_{EE}+0.8 V only p-n-p pair Q_3, Q_4 operates.
2. From the positive supply voltage V_{CC} to V_{CC}-0.8 V only n-p-n pair Q_1, Q_2 operates.
3. In the intermediate range, both pairs are operates.

The voltage drop over resistors R_8-R_{11} is usually in the order of 100 mV, so the actual CM-input range extends beyond both supply rails for about 400 mV. This is not depicted in Fig. 2.2 for reasons of simplicity, but it is nevertheless an important feature of the input stage, especially if the OpAmp is used as an inverting amplifier; a small input signal below the negative supply still yields a positive output voltage.

If the CM-input voltage moves from one range into another, the sum of the tail currents flowing through the input transistors changes by a factor of two and consequently so does the transconductance of this input stage. The overall bandwidth of the OpAmp is directly proportional to this transconductance and therefore an optimal frequency compensation is difficult to obtain. Moreover, signal distortion occurs if one of the current sources I_{B1} or I_{B2} saturates and therefore the need arises for an input stage with the same rail-to-rail capabilities, but with a more constant transconductance and with precautions against the saturation of the current sources.

2.2 COMPLEMENTARY INPUT STAGE WITH CONSTANT TRANSCONDUCTANCE

The drawbacks of the input stage described in the previous Section can be overcome if, at any CM-input voltage, the sum of the tail currents of the two input pairs is kept constant. Because the transconductance of each input pair is directly proportional to its tail current, the transconductance is then also constant. A circuit that does just that is shown in Fig. 2.3 [2], [3]. If the CM-input voltage is near the negative supply rail, current source I_{B1} activates the p-n-p pair Q_3, Q_4,

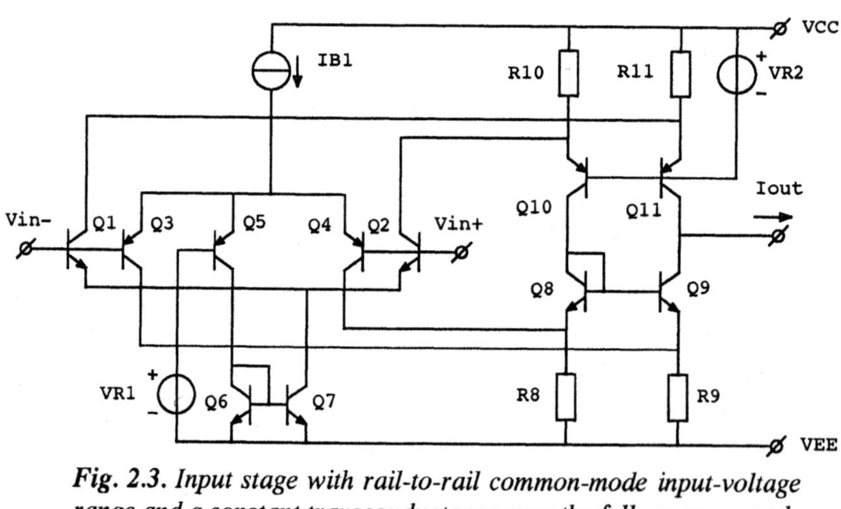

Fig. 2.3. Input stage with rail-to-rail common-mode input-voltage range and a constant transconductance over the full common-mode range.

provided that the supply voltage is high enough. This p-n-p pair is now able to handle the input signal. If the CM-input voltage is now raised above the reference voltage V_{R1}, transistor Q_5 takes away the current from current source I_{B1} and through Q_6 and Q_7 supplies the n-p-n pair Q_1, Q_2. The p-n-p pair Q_3, Q_4 is now switched off and the signal operation is performed by the n-p-n pair Q_1, Q_2. Even in the turn-over range of Q_5, the sum of the tail currents of the input pairs, and therefore also the total transconductance of the input stage, is kept constant. The value of the reference voltage source V_{R1} is 0.8 V, high enough to enable the n-p-n pair to function properly when the CM-input voltage exceeds this value. The transistors Q_8-Q_{11} and resistors R_8-R_{11} sum the collector currents of the input pairs, just as depicted in Fig. 2.1.

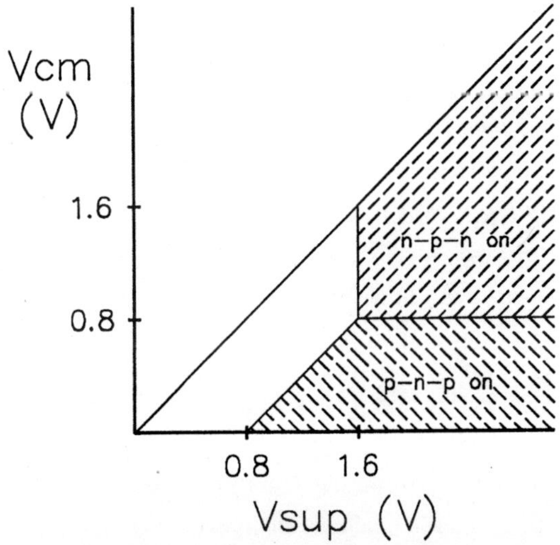

Fig. 2.4. The operational regions of the input stage shown in Fig. 2.3. Only one of the input pairs is active.

Fig. 2.4 shows the operational regions of the input stage shown in Fig. 2.3; comparing this figure with Fig. 2.2 it can be seen that now either the p-n-p pair or the n-p-n pair is active, but not both. For supply voltages below 1.6 V (the sum of V_{R1}, the base-emitter voltage of Q_5 and the saturation voltage of I_{B1}) the n-p-n pair is completely shut off and only the p-n-p pair is operative. Again, although not shown in Fig. 2.4, the CM-input range exceeds the supply-voltage rails for about 400 mV, just as it did with respect to the circuit shown in Fig. 2.1.

If the influence of additive interference signals on the behavior of the OpAmp is investigated, it is assumed that only the input stage should be taken in account. This assumption is valid if the voltage and current gain of the input stage are large. If the cascode Q_8-Q_{11}, R_8-R_{11} is assumed not to degrade the performance of the input stage, the transconductance g_i of the input stage is:

$$g_i = g_{1-4} = \frac{I_{B1}}{2V_T} \tag{2.1}$$

and the voltage gain is given by:

$$A_V = \frac{1}{2} g_{1-4} R_L \tag{2.2}$$

where g_{1-4} is the transconductance of one of the transistors Q_1-Q_4, and R_L is the input resistance of the intermediate stage. If the tail-current source I_{B1} in Fig. 2.3 is 10 μA, the transconductance g_{1-4} is 0.2 mmho, and a typical value of the input resistance of the intermediate stage is 10 MΩ. The voltage gain of the input stage is then 2.10^3, high enough to neglect the influence of the next stages. The current gain of the input stage depends on which of the two input pairs is operative:

$$A_i = \beta_{1-4} \tag{2.3}$$

Even a relatively small p-n-p current gain in the order of 20 to 60, is usually large enough to ignore the next stages in the analyses of interference signals.

First the offset voltage V_{OS} of the input stage shown in Fig. 2.3 is examined. The influence of the mismatch in the base currents of the transistors Q_8-Q_{11} is neglected and the voltage drop over the resistors R_8-R_{11} is supposed to be great compared to the thermal voltage V_T. Further, the current through Q_8-Q_{11} is assumed to be in the same order as the current through Q_1, Q_2 and Q_3, Q_4. If the n-p-n input pair Q_1, Q_2 is conducting, the input offset voltage can be calculated:

$$V_{OS} = V_T \left(\frac{\Delta R_{8,9}}{R_{8,9}} + 2\frac{\Delta R_{10,11}}{R_{10,11}} + \frac{\Delta I_{S1,2}}{I_{S1,2}} \right) \tag{2.4}$$

Similarly, if the p-n-p pair Q_3, Q_4 is conducting, the offset voltage is given by:

$$V_{OS} = V_T \left(2\frac{\Delta R_{8,9}}{R_{8,9}} + \frac{\Delta R_{10,11}}{R_{10,11}} + \frac{\Delta I_{S3,4}}{I_{S3,4}} \right) \tag{2.5}$$

With a resistor mismatch of 1% and a mismatch of 5% in the I_S, the saturation current of the input transistors, the offset voltage is in the order of 2 mV, but when the input transistors and the resistors R_8-R_{11} are used in a cross-coupled layout the mismatch in the resistors and in the I_S is typically 0.2% and 1% respectively, and the resulting offset can be kept below 0.4 mV [4].

The temperature coefficient of the offset voltage, or offset drift is given by:

$$\frac{dV_{OS}}{dT} = \frac{V_{OS}}{T} \tag{2.6}$$

because the thermal voltage V_T is equal to kT/q and the other factors in (2.4) are considered temperature independent. At room temperature the drift is in the order of 1-5 μV/°C.

The input bias current is naturally dependent on the value of the tail current source I_{B1} and on the value of the current gain β of the input transistors Q_1-Q_4. The direction of the input bias current changes as the CM-input voltage crosses the reference voltage V_{R1} and the signal operation is transferred from the p-n-p pair to the n-p-n pair. To reduce the influence of this undesirable property, the tail current I_{B1} is kept as small as possible. In practice, values range from 2-10 μA, causing an input bias current in the order of 20-500 nA.

The input offset current of the n-p-n pair can be calculated in the same way the as the offset voltage:

$$I_{OS} = \frac{I_{C1,2}}{\beta_{1,2}} \left(3\frac{\Delta R_{8-11}}{R_{8-11}} + \frac{\Delta\beta_{1,2}}{\beta_{1,2}} \right) \tag{2.7}$$

with $I_{C1,2}$ the collector current of $Q_{1,2}$. The p-n-p input offset current is:

$$I_{OS} = \frac{I_{C3,4}}{\beta_{3,4}} \left(3\frac{\Delta R_{8-11}}{R_{8-11}} + \frac{\Delta\beta_{3,4}}{\beta_{3,4}} \right) \tag{2.8}$$

If a cross-coupled or common-centroid geometry is again assumed, the mismatch in the resistor value is in the order of 0.2% and in the current gain β of the transistors it is 5%. This yields an input offset current of 1-25 nA.

The common-mode rejection ratio (*CMRR*) of the input stage is defined as:

$$CMRR = \frac{\Delta V_{CM}}{\Delta V_{OS}} \tag{2.9}$$

The offset voltage of the input stage shown in Fig. 2.3 changes between that of the n-p-n input pair and that of the p-n-p input pair. Eqs. (2.4) and (2.5) show that the offset-voltage is in the order of 0.5 mV. The CM range in which this turnover takes place is determined by current switch Q_5. If this turnover range is 200 mV, the *CMRR* is 52 dB. Calculated over the full CM-input range however, the *CMRR* is much larger; with a supply voltage of 5 V, the *CMRR* is 80 dB. This illustrates that it is necessary to be explicit in the definition that is being used.

The noise performance of the input stage is examined by calculating the contributions of the circuit elements. The equivalent output noise caused by Q_{10}, Q_{11}, R_{10} and R_{11} is [5]:

$$\overline{i^2_{no10,11}} = \left(\frac{\frac{1}{2}}{R_{10,11} + r_{e10,11}} \right)^2 4kT\left(\frac{1}{2}r_{e10} + r_{b10} + \frac{1}{2}r_{e11} + r_{b11} + R_{10} + R_{11}\right) \Delta f \tag{2.10}$$

where r_e is the emitter resistance and r_b is the parasitic base resistance. At current levels of 1-10 μA the influence of $r_{b1,2}$ can be neglected, and if further the voltage over the resistors $R_{8\text{-}11}$ is again large compared to the thermal voltage V_T, or in other words $R_{8\text{-}11}$ is large compared to $r_{e8\text{-}11}$ then (2.10) becomes:

$$\overline{i^2_{no10,11}} = 4kT\frac{1}{2R_{10,11}}\Delta f \tag{2.11}$$

Equally, for Q_8, Q_9, R_8 and R_9:

$$\overline{i^2_{no8,9}} = 4kT\frac{1}{2R_{8,9}}\Delta f \tag{2.12}$$

If the n-p-n input pair Q_1, Q_2 is conducting, the contribution of $Q_{8\text{-}11}$ and $R_{8\text{-}11}$ to the input noise then is:

$$\overline{u^2_{ni8-11}} = (2r_{e1,2})^2 4kT\frac{1}{R_{8-11}}\Delta f \tag{2.13}$$

The input noise caused by Q_1, Q_2 is:

$$\overline{u^2_{ni1,2}} = 4kT(r_{e1,2}+2r_{b1,2})\Delta f \tag{2.14}$$

The influence of $r_{b1,2}$ is again ignored, so the total equivalent input noise voltage is:

$$\overline{u^2_{ni\ tot}} = 4kTr_{e1,2}\left(1+\frac{4r_{e1,2}}{R_{8-11}}\right)\Delta f \tag{2.15}$$

If the p-n-p input pair is operative, $r_{e1,2}$ is replaced by $r_{e3,4}$, but since both have the same value the total equivalent input noise voltage does not change. When, for instance, the tail current I_{B1} is 5 μA and $R_{8\text{-}11}$ are 25 kΩ, the total input noise voltage is 20 nV/√Hz.

Neglecting the 1/f noise, the equivalent input noise current at both input terminals of the input stage is dominated by the contribution of the base current of Q_1, Q_2 or Q_3, Q_4:

$$\overline{i_{ni\ tot}^2} = 4kT\left(\frac{1}{2r_{be1-4}}\right)\Delta f \tag{2.16}$$

At a tail current of 5 μA again, and with a current gain $\beta_{1,2}$ of 100, the input noise current is only 0.12 pA/√Hz so it can usually be disregarded.

The rail-to-rail input stage described in this Section consists of relatively few circuit parts and it has good specifications; it is therefore very well suited for use in low-voltage OpAmps.

2.3 RAIL-TO-RAIL INPUT STAGE AT 1 VOLT

The input stage shown in Fig. 2.3 already has a rail-to-rail CM input-voltage range if the supply voltage is equal to or higher than 1.6 V. At a supply voltage of 1 V, the p-n-p input pair shown in Fig. 2.3 can only operate near the negative supply rail and the n-p-n pair cannot function at all. Nevertheless, there may be applications where there is a need for the rail-to-rail capability at a supply voltage of only 1 V. If the OpAmp is used as a voltage follower for instance, both the input and the output voltage should be as large as possible, and the input and output-voltage ranges certainly should include the middle of the supply-voltage range. Another advantage of a rail-to-rail input-voltage range is that it allows a great degree of freedom to the designer who uses the OpAmp in a low-voltage system.

To overcome this problem, level-shift resistors R_{L1}-R_{L4} have been inserted between the input terminals and the base of the n-p-n and p-n-p input pairs, Fig. 2.5 [6]. The level-shift resistors are activated by the current sources I_{L1}-I_{L4} and in this way either the n-p-n or the p-n-p pair is enabled to operate. With the level shift activated, however, the input terminals can no longer reach the positive and negative supply rails. Therefore the current sources I_{L1}-I_{L4} are made dependent on the CM-input voltage in such a way that the current, and therefore the level-

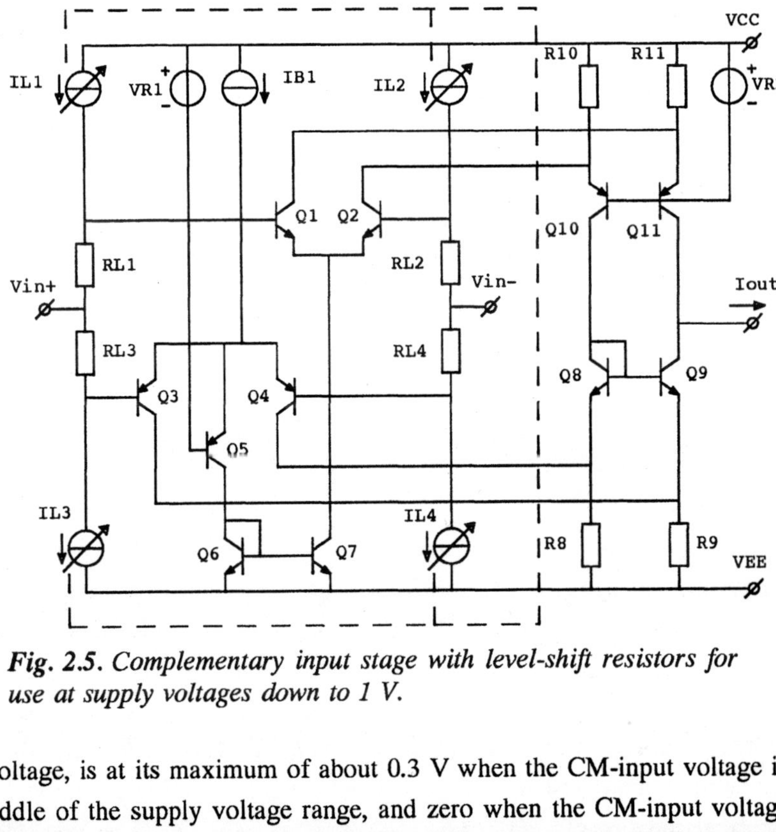

Fig. 2.5. Complementary input stage with level-shift resistors for use at supply voltages down to 1 V.

shift voltage, is at its maximum of about 0.3 V when the CM-input voltage is in the middle of the supply voltage range, and zero when the CM-input voltage is near either one of the supply rails. The level-shift voltage then should be of the shape shown in Fig. 2.6. The p-n-p input pair Q_3-Q_4 in Fig. 2.5 operates on the input signals in the lower half of the CM-input range and n-p-n pair Q_1-Q_2 deals with signals in the upper half. The n-p-n pair is now able to function, because the reference voltage V_{R1} of approximately 0.7 V is connected to the positive supply rail instead of to the negative supply rail. The voltage over current source I_{B1} therefore still is about 0.1 V when the n-p-n pair tends to become active. The transistors Q_8-Q_{11} and the resistors R_8-R_{11} again make up the summing circuit that adds the collector currents of the input pairs.

Fig. 2.7 shows an implementation of the circuit depicted in Fig. 2.5. The level-

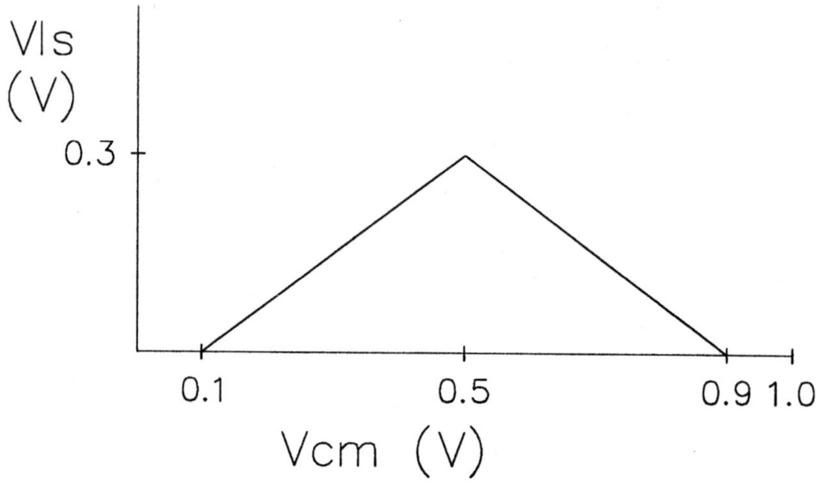

Fig. 2.6. The level-shift voltage $I_{L1\text{-}4}R_{L1\text{-}4}$ as function of the common-mode input voltage.

shift current sources I_{L1}-I_{L4} are realized by transistors Q_{21}-Q_{24} and transistors Q_{12}-Q_{20} form the dynamic level-shift generator that controls the current sources Q_{21}-Q_{24} in a way that yields the dependency shown in Fig. 2.6. Differential amplifier Q_{12}-Q_{13} compares the voltage across the tail-current source I_{B1} with the reference voltage V_{R2} and the differential amplifier Q_{14}-Q_{15} compares the voltage over the current source Q_7 with reference voltage V_{R3}. Both differential amplifiers transfer just enough current from current source I_{B2} (via current mirror Q_{16}, Q_{17} and current mirrors Q_{18}, Q_{19}, Q_{21}, Q_{22} and Q_{20}, Q_{23}, Q_{24}) to the level-shift resistors R_{L1}-R_{L4} to keep at least one of the tail current sources I_{B1} or Q_7 from saturating.

To further illustrate the functioning of the circuit shown in Fig. 2.7, suppose that the CM-input voltage is 0.0 V and the supply voltage is 1.0 V. The base-emitter voltage of conducting input transistors is assumed to be 0.6 V and the reference voltage sources V_{R1}, V_{R2} and V_{R3} have the values 0.7 V, 0.2 V and 0.2 V, respectively. At a CM-input voltage of 0.0 V, the n-p-n input transistors Q_1, Q_2 are unable to function and are switched off. Current source I_{B3}, only about 100 nA, pulls the base of transistor Q_{14} down to the negative supply rail. Transistor

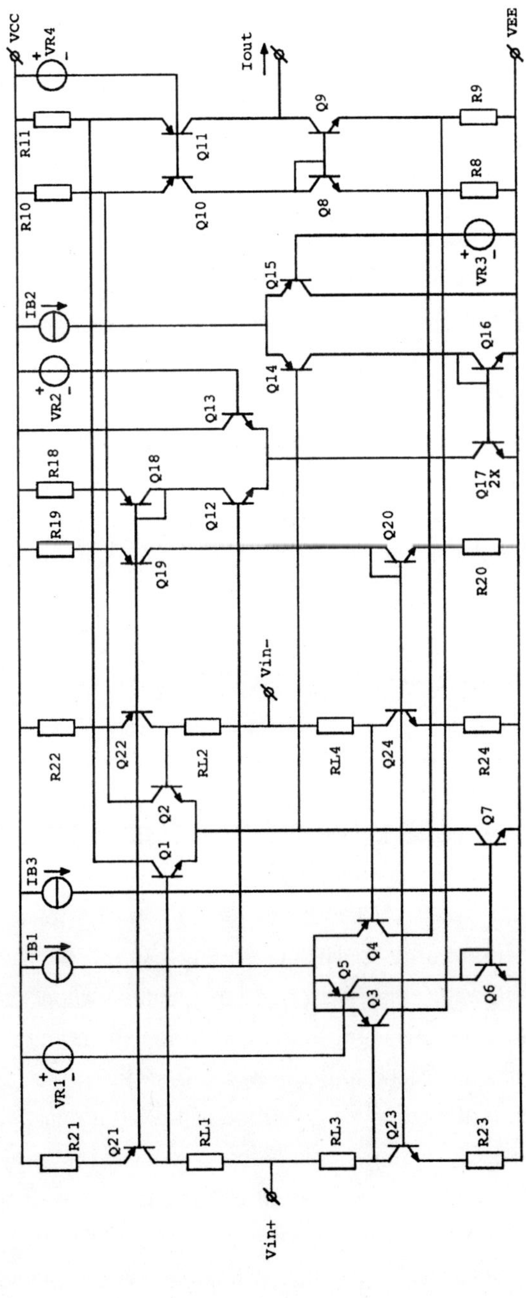

Fig. 2.7. Complementary input stage with dynamic level-shift generator.

Q_{14} is conducting and transfers the tail current I_{B2} to the differential amplifier Q_{12}, Q_{13}. The voltage at the emitters of the p-n-p input pair is 0.6 V and in this case the differential amplifier Q_{12}, Q_{13} directs its tail current to the positive supply; the level shift is not activated. When the CM-input voltage rises to 0.10 V, the voltage at the emitters of the p-n-p input pair is 0.7 V and the voltage difference between the bases of differential amplifier Q_{12}, Q_{13} is 0.1 V. Now transistor Q_{12} starts to take over a part of the current through Q_{13} and mirrors it to the level-shift resistors. While the CM-input voltage is further rising, the negative-feedback loop consisting of the p-n-p input stage Q_3, Q_4, the differential amplifier Q_{12}, Q_{13} and the current mirrors Q_{18}-Q_{24} keeps the voltage difference at the input of the differential amplifier Q_{12}, Q_{13} small. Hence the emitter, and therefore the base voltages of p-n-p input stage Q_3, Q_4 remain more or less constant when the CM-input voltage is further increased. Thus, the voltage over the level-shift resistors R_{L3}, R_{L4} is forced to increase by an increase of the level-shift currents. The voltage over the level-shift resistors R_{L1}, R_{L2} also increases and eventually the base voltage of the n-p-n input-pair transistors Q_1, Q_2 is high enough to enable these transistors to function. The CM-input voltage is now halfway between the positive and negative supply voltages and both complementary input stages are conducting. At this point the level-shift current is at its maximum and is also controlled through the emitters of the n-p-n input pair by the differential amplifier Q_{14}, Q_{15}. When the CM-input voltage is further increased, the differential amplifier Q_{14}, Q_{15} fully takes control of the feedback loop and the level-shift current is decreased. Finally, when the CM-input voltage reaches 0.90 V, the voltage at the emitters of the n-p-n input pair Q_1, Q_2 is 0.3 V and hence the voltage difference between the bases of the differential-amplifier transistors Q_{14}, Q_{15} is 0.1 V. The current I_{B2} now completely flows through transistor Q_{15} to the negative supply and the level shift is turned off. When the level-shift current is near its maximum, the current switch Q_5 gradually takes away the tail current I_{B1} from the p-n-p input pair and steers it through the current mirror Q_6, Q_7 to the n-p-n input pair.

An increase in the supply voltage automatically reduces the voltage range in which the level-shift is active. This is a result of the connection of the reference voltage source V_{R2} to the positive supply rail and reference voltage V_{R3} to the negative supply rail. When the CM-input voltage at which the level shift is

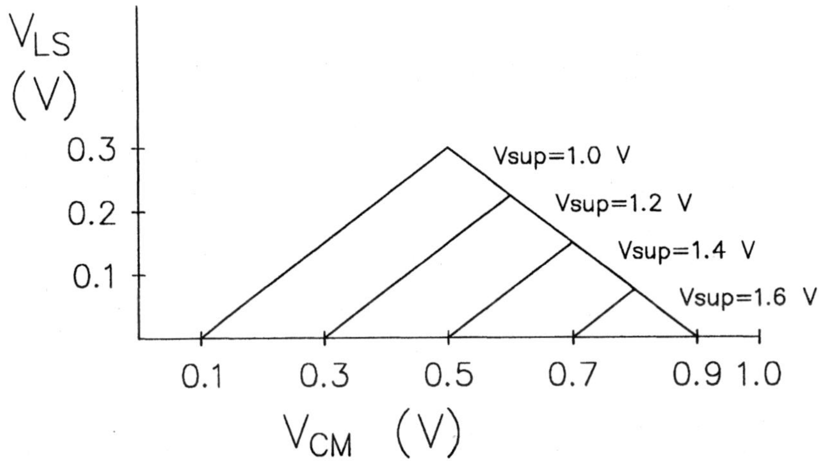

Fig. 2.8. Curves of the level-shift voltage as function of the common-mode input voltage. Chip temperature is 27 °C; V_{sup} = 1.0, 1.2, 1.4, and 1.6 V from top to bottom.

switched on by Q_{12}, Q_{13} equals the CM-input voltage at which it is switched off by Q_{14}, Q_{15}, the level shift is not activated at all. This happens at a supply voltage of V_{sup} = 0.2+0.1+0.6-(-0.2-0.1-0.6) = 1.8 V. Fig. 2.8 shows this dependency on the supply voltage.

The chip temperature is also taken automatically into account, because the CM-input voltage is sensed at the emitters of the p-n-p and n-p-n input-pair transistors. If, for instance, the temperature is high, the range in which the complementary input stages require the level shift to function is smaller than it is at room temperature. This is shown in Fig. 2.9.

Fig. 2.10 shows the operational regions of the 1-V rail-to-rail input stage. The use of the level shift extends the region in which the OpAmp is able to function from rail to rail down to a supply voltage of 1 Volt. Below 1 Volt either the p-n-p or the n-p-n input pair is able to operate, but not both.

The offset voltage of the input stage shown in Fig. 2.7 can be analyzed by inspecting the contributions of the various circuit parts. The offset of the input pairs Q_1, Q_2 and Q_3, Q_4 and of the sum-circuit Q_8-Q_{11} and R_8-R_{11} has already

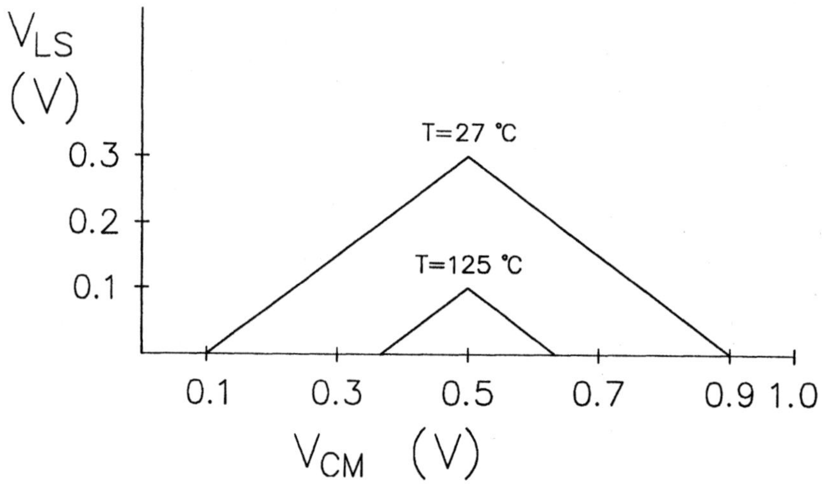

Fig. 2.9. Curves of the level-shift voltage as function of the common-mode input voltage. V_{sup} = 1.0 V; chip temperature is 27 °C and 125 °C from top to bottom, respectively.

been analyzed in Section 2.2 and was given by (2.4). Under the same assumptions, the contribution of the input pairs is again given by:

$$V_{OS1} = V_T \left(3 \frac{\Delta R_{8-11}}{R_{8-11}} + \frac{\Delta I_{S1-4}}{I_{S1-4}} \right) \tag{2.17}$$

With a resistor mismatch of 0.2% and a transistor mismatch of 1%, the resulting offset contribution is 0.4 mV. If the level-shift current is at its peak value I_{Lmax}, the contribution of the mismatch in the level-shift resistors R_{L1}-R_{L4} is at its maximal value:

$$V_{OS2} = R_{L1-4} I_{Lmax} \left(\frac{\Delta R_{L1-4}}{R_{L1-4}} \right) \tag{2.18}$$

With a maximum level-shift voltage $I_{Lmax}R_{L1\text{-}4}$ of 0.3 V, the level-shift resistors contribute 0.6 mV to the total offset voltage. The offset voltage caused by

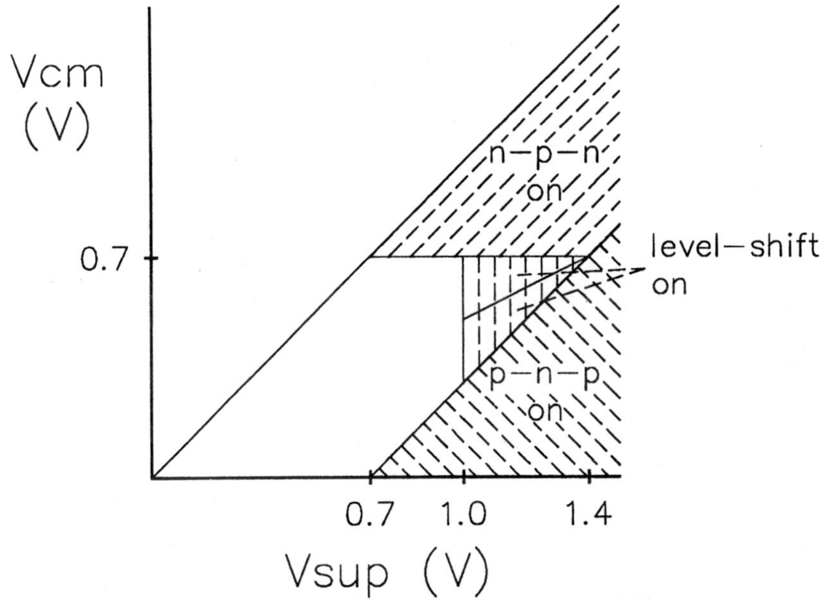

Fig. 2.10. *The operational regions of the input stage shown in Fig. 2.7.*

mismatch in the transistors Q_{20}-Q_{24} and in the resistors R_{20}-R_{24} is:

$$V_{OS3} = R_{L1-4} I_{Lmax} \left(\frac{r_{e20-24}}{r_{e20-24} + R_{20-24}} \frac{\Delta I_{S20-24}}{I_{S20-24}} + \right. \\ \left. + \frac{R_{20-24}}{r_{e20-24} + R_{20-24}} \frac{\Delta R_{20-24}}{R_{20-24}} \right) \tag{2.19}$$

If the resistors $R_{20\text{-}24}$ are 15 kΩ and the maximum level-shift current is 5 μA, V_{OS3} is 1.2 mV. The total offset of the OpAmp is the sum of V_{OS1}, V_{OS2} and V_{OS3} which is equal to 2.2 mV.

The offset drift can be calculated by differentiating the (2.17), (2.18) and (2.19). At room temperature, the resulting values are 1.3 μV/°C, 4 μV/°V and 8 μV/°C, respectively, giving a total of about 13 μV/°C.

With respect to the input bias current, the same considerations are valid that were stated in Section 2.2. Additionally, if the level shift is active, the base currents of transistors Q_{20}, Q_{23} and Q_{24} also contribute to the input bias current. Practical values for I_{B1}=5 μA range from 100-300 nA.

The input offset current is dominated by the current caused by the mismatch in the level-shift current sources and by the mismatch in the current gain β of the input pairs. The input offset current of the n-p-n input pair is:

$$I_{OS} = \frac{I_{C1,2}}{\beta_{1,2}} \frac{\Delta\beta_{1,2}}{\beta_{1,2}} + I_{Lmax} \left(\frac{r_{e21,22}}{r_{e21,22}+R_{21,22}} \frac{\Delta I_{S21,22}}{I_{S21,22}} + \frac{R_{21,22}}{r_{e21,22}+R_{21,22}} \frac{\Delta R_{21,22}}{R_{21,22}} \right) \tag{2.20}$$

and for the p-n-p input pair it is:

$$I_{OS} = \frac{I_{C3,4}}{\beta_{3,4}} \frac{\Delta\beta_{3,4}}{\beta_{3,4}} + I_{Lmax} \left(\frac{r_{e23,24}}{r_{e23,24}+R_{23,24}} \frac{\Delta I_{S23,24}}{I_{S23,24}} + \frac{R_{23,24}}{r_{e23,24}+R_{23,24}} \frac{\Delta R_{23,24}}{R_{23,24}} \right) \tag{2.21}$$

If the same conditions are met that were assumed in calculating the offset voltage and further the current gain β is 100 for the n-p-n transistors, 30 for p-n-p's and the mismatch in β is 5%, the value of the input offset current is in the order of 15-50 nA.

The common-mode rejection ratio (*CMRR*) is determined by the offset voltage change if the signal operation is taken over from the p-n-p to the n-p-n input pair. With the figures given above, this change can be at the most 4.2 mV, giving a worst-case *CMRR* of 34 dB. If the supply voltage is increased, the level shift is de-activated and only the offset change of the two input pairs remains; for a supply voltage of 5 V the expected *CMRR* is 75 dB.

The equivalent input noise voltage consists of three components: the noise voltage caused by the actual input pairs and summing circuit, the voltage attributable to the level-shift resistors and the noise voltage resulting from the level-shift current sources. The contribution of the n-p-n input pair Q_1, Q_2, provided that that is conducting, is given in (2.15):

$$\overline{u_{ni1}^2} = 4kTr_{e1-4}\left(1+\frac{4r_{e1-4}}{R_{8-11}}\right)\Delta f \tag{2.22}$$

The input noise voltage caused to the level-shift resistors is:

$$\overline{u_{ni2}^2} = 4kT\left(2R_{L1-4}\right)\Delta f \tag{2.23}$$

The contribution of the level-shift current sources Q_{21}-Q_{24} and their emitter resistors is given by:

$$\overline{u_{ni3}^2} = 4kT\left(\frac{\left(R_{L1-4}\right)^2}{\left(r_{e21-24}+R_{21-24}\right)^2}\left(r_{e21-24}+2R_{21-24}\right)\right)\Delta f \tag{2.24}$$

The total equivalent input noise is the sum of (2.22), (2.23) and (2.24). With I_{B1}=5 μA, I_{LS}=5 μA, R_{L1-4}=60 kΩ, R_{21-24}=15 kΩ, R_{8-11}=25 kΩ, $\beta_{1,2}$=100 and $\beta_{3,4}$=30 the input noise voltage is $\sqrt{\overline{u_{ni1}^2}+\overline{u_{ni2}^2}+\overline{u_{ni3}^2}}$ = 87 nV/√Hz. If the level shift is not active, the third factor in the sum disappears, and the input noise voltage decreases to 50 nV/√Hz.

The equivalent input noise current is:

$$\overline{i_{ni}^2} = 4kT\left(\frac{R_{21-24}}{\left(r_{e21-24}+R_{21-24}\right)^2}+\frac{1}{2r_{be1-4}}\right)\Delta f \tag{2.25}$$

When the figures given above are used, the maximum input noise current is 0.8 pA/√Hz if the level-shift current is at its maximum and 0.23 pA/√Hz if the level shift is switched off.

Comparing the input stage introduced in this Section with the one shown in Fig. 2.3, we found that the extension of rail-to-rail operation down to 1 Volt is exchanged for a deterioration of such characteristics as noise performance and offset. Although this disadvantage mostly disappears if the level shift is not active, it implies that this input stage will only be used when the need for rail-to-rail operation at very low supply voltages is clearly apparent, or when optimal flexibility in the use of the opamp is demanded.

2.4 INPUT STAGE WITH EXTENDED INPUT-VOLTAGE RANGE

The input stages described in Sections 2.1, 2.2 and 2.3 have CM-input voltage ranges that extend beyond the supply rails for approximately 400 mV. This feature is very useful when the OpAmp is used as an inverting amplifier; a small input signal below the negative supply still yields a positive output voltage. However, if the CM-input voltage is still further increased or decreased, the transistors of the input pairs become saturated. In the case of low-voltage OpAmps especially, it is

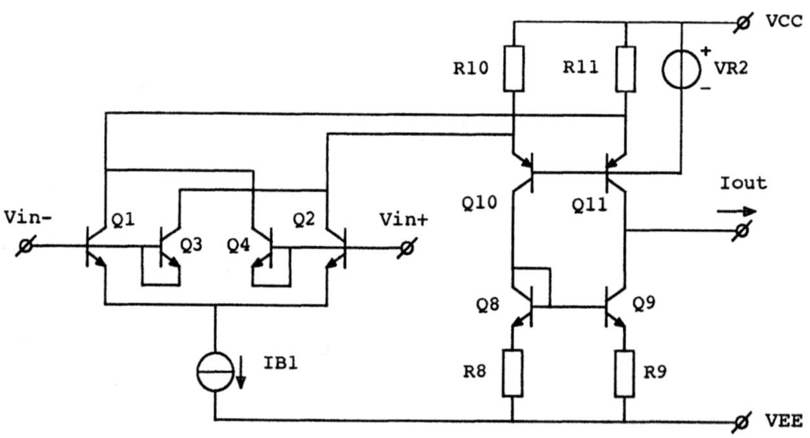

Fig. 2.11. Input stage with diodes to prevent reversal of the output signal.

quite possible that the CM-input voltage is unintentionally driven beyond these specification limits. What happens then is that, because of this saturation, the collector currents of the input-pair transistors reverse. This reversal is then transferred through the amplifier, leading eventually to reversal of the output signal of the OpAmp. This Section describes the means to prevent this.

Fig 2.11 shows an input stage that consists of n-p-n input pair Q_1, Q_2 and summing circuit Q_8-Q_{11}, equal to the input stages shown in the previous Sections. For reasons of simplicity, only the n-p-n input pair is shown but the addition to the circuit, which is described below, is valid both for n-p-n as well as for p-n-p input pairs. If the voltage drop over the resistors R_{10}, R_{11} is assumed to be 100 mV, and there exists a small positive voltage difference between the bases of Q_1 and Q_2, transistor Q_1 starts to saturate when the CM-input voltage is raised to 400 mV above the positive supply rail. The base-collector diode of Q_1 then becomes forward biased and the current that is normally pulled out of resistor R_{11} is decreased with the base-to-collector current of the saturating input transistor. To prevent the resulting reversal of the output signal, transistors Q_3 and Q_4 are added to the circuit shown in Fig. 2.11. These transistors have twice the emitter area of the transistors of the differential input pair, and therefore the base-collector current of Q_3, if the input stage becomes saturated, is also twice as large as that of Q_1. This larger saturation current is then used to over-compensate the saturation current of Q_1, by electrically cross-coupling the sensing transistors Q_3, Q_4 with the input-pair transistors Q_1, Q_2. Thus the signal polarity remains correct. This results in the ability to overdrive the inputs of the circuit to a CM voltage of at least one diode voltage beyond the supply rails without unintentional reversal of the output voltage. To prevent destruction of Q_3 and Q_4 if the CM-input voltage is further increased, the total OpAmp is normally equipped with clipping diodes.

Transistors Q_3, Q_4 in Fig. 2.11 have twice the emitter area of the input-pair transistors Q_1, Q_2, but, in fact, any ratio of the two emitter area's greater than one is sufficient. Another way to accomplish the over-compensation of the saturation current of Q_1, Q_2 is by inserting resistors in the collector leads of these transistors; if the sensing diodes, which are now scaled 1:1 with Q_1, Q_2, are directly coupled to the summing circuit, their saturation current is then larger than that of the input transistors. The CM-input range is then slightly reduced, and therefore the circuit shown in Fig. 2.11 is preferred.

Naturally, the input bias current of the circuit depicted in Fig. 2.11 increases, and also the input-offset voltage and current and the CMRR, degenerate to values that are less favorable than the ones calculated in Sections 2.2. However, the addition of Q_3, Q_4 hardly has any influence on the specifications of the input stage when the CM-input voltage is in the normal operational region, and Q_3, Q_4 are not conducting.

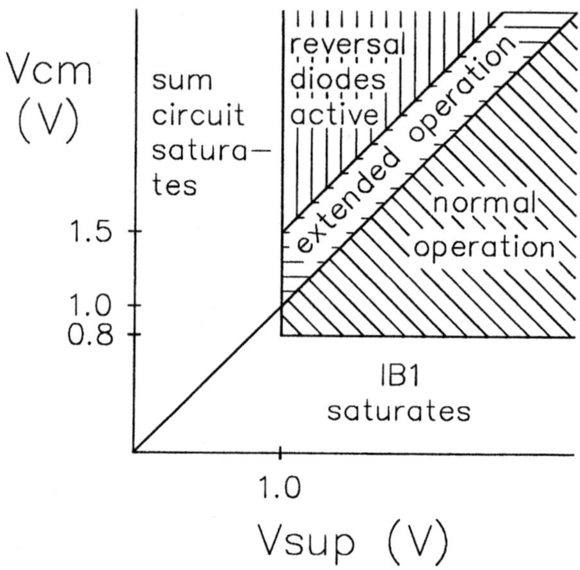

Fig. 2.12. Operational regions of the input stage depicted in Fig. 2.11.

Fig. 2.12 shows the operational regions of the n-p-n input stage shown in Fig. 2.11. Note once more that the circuit depicted in Fig. 2.11 can be extended by the addition of a similar p-n-p input pair, also with its own p-n-p sensing transistors. The operational regions of Fig. 2.12 would then be larger than they are now. The purpose of Fig. 2.12 is firstly to show the extension beyond the supply rail that should have been present also in Figs. 2.2, 2.4 and 2.10 but were omitted there for reasons of simplicity, and, secondly, to illustrate the further addition to the operational regions that originate from the insertion of the sensing transistors.

According to Fig. 2.12, the minimum supply voltage of the circuit shown in Fig. 2.11 is 1 Volt. This is because of the saturation of transistor Q_{10}. If the base current of Q_8, Q_9 is supplied in a different way, and Q_8 is not a diode, the minimum value of the supply voltage can be slightly lower than 1 Volt. If a p-n-p input pair was added to the circuit shown in Fig. 2.11, the area indicated by "I_{B1} saturates" would also be included in the operational regions. In [7] an input stage is described that allows the input voltage to be even further beyond the supply rails.

2.5 REFERENCES

[1] B. Blauschild, "Differential amplifier with rail-to-rail input capability and controlled transconductance", U.S. Pat. Appl. Ser. No. 602.231, filed April 19, 1983.

[2] J. H. Huijsing and R. J. van de Plassche, "Differential amplifier with rail-to-rail input capability and controlled transconductance", U.S. Pat. Appl. Ser. No. 602.231, filed April 19, 1984.

[3] J. H. Huijsing and D. Linebarger, "Low-Voltage Operational Amplifier with Rail-to-Rail Input and Output Ranges", *IEEE J. Solid-State Circuits*, Vol. SC-20, pp. 1144-1150, Dec. 1985.

[4] J. H. Huijsing, *Integrated Circuits for Accurate Linear Analogue Electrical Signal Processing*. Ph.D. Thesis, Delft Univ. Technol., Delft, the Netherlands 1981, p. 65.

[5] P. R. Gray and R. G. Meyer, *Analyses and Design of Analog Integrated Circuits*. New York: Wiley, 1984, pp. 681-686.

[6] J. Fonderie, M. M. Maris, E. J. Schnitger and J. H. Huijsing, "1-V Operational Amplifier with Rail-to-Rail Input and Output Ranges", *IEEE J. Solid-State Circuits*, Vol. SC-24, pp. 1551-1559, Dec. 1989.

[7] G. van der Horn and J. H. Huijsing, "Extension of the Common-Mode Range beyond the Supply Rails of Operational Amplifiers and Comparators", in *Proc. ESSCIRC*, 1992, pp. 187-190.

CHAPTER 3
OUTPUT STAGES

The output stage of a low-voltage OpAmp should be able to deliver an output voltage signal that is as large as possible, preferably from one supply-voltage rail to the other. After all, the supply-voltage range is by definition not very great so we want to be able to make full use of it. This rules out the use of emitter followers as output transistors, because such a use would result in the loss of one diode voltage. The actual output transistors of the output stages discussed in this Chapter therefore have a common-emitter configuration.

When dealing with low-voltage OpAmps, it is of course of major interest to know the lowest supply voltage at which the circuit is still able to function. Two other DC characteristics, the large-signal current gain and the total amount of current drawn from the supplies, related to the output current, also determine the suitability of each output stage. These are the first items that will be investigated in each of the following Sections.

Thereafter, the discussion will focus on the small-signal and high-frequency behavior of the output stages. Initially, current-driven output stages with collector outputs have two dominant pole frequencies, one at the input and one at the output. Because the output stage is part of the negative feedback loops within the OpAmp and probably also around it, these two poles will give rise to oscillations. Therefore Miller capacitors are inserted between the input and the output. They split the poles, leaving the output stage with one dominating pole, which yields a stable frequency response [1]. The output stages discussed in this Chapter are shown with a large capacitive load and no resistive load, because the latter is the most difficult condition where the compensation of the circuits is concerned.

Fig. 3.1 shows the most elementary output stage of those described in this Chapter. The output stage has two complementary parts, a p-n-p transistor connected to the positive supply rail and an n-p-n transistor connected to the

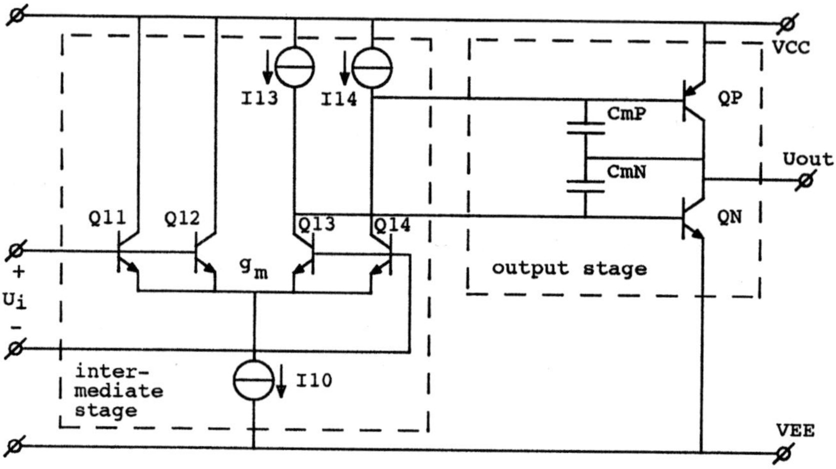

Fig. 3.1. Simple complementary output stage, preceded by an intermediate stage with transconductance g_m.

negative supply rail. Thus the entire supply-voltage range is included except for the small saturation voltages of the output transistors. The discussion of the output stages will not include the operation of the class-AB current control, nor that of circuitry for current limitation and protection against deep saturation of the output transistors, which is discussed in Chapter 4.

Fig. 3.1 also shows a stage preceding the output stage. This stage, the intermediate stage, transfers its input voltage to an output current, which in turn is the input current of the output stage. Although the intermediate stage is not showed with the output stages in the following Sections, its output is represented by a current source. In the small-signal equivalent circuits, this current source then has the value of $g_m U_i$, the product of the input voltage of the intermediate stage and its transconductance. The reason for this is that this transconductance directly controls the unity-gain bandwidth of the combination of the two stages.

Only one of the complementary parts of the output stage, the n-p-n part will be shown in the schematics in the following Sections. The p-n-p side is completely complementary however, and examples of both n-p-n and p-n-p output stages are given in the text.

In the first Section, the simple common-emitter stage, as seen in Fig. 3.1 is analyzed. Then we examine the popular Darlington output stage and the third Section deals with the Widlar output stage. The last two Sections of this Chapter describe two new principles that try to bypass the trade-off between achieving a high gain and a high bandwidth, a problem that is discussed in the first three Sections.

3.1 COMMON-EMITTER OUTPUT STAGE

The most simple output stage that can be used in a low-voltage OpAmp is one transistor in common-emitter (CE) configuration. Fig. 3.2 shows the circuit diagram of the n-p-n CE output stage. The current source $g_m U_i$ drives the output

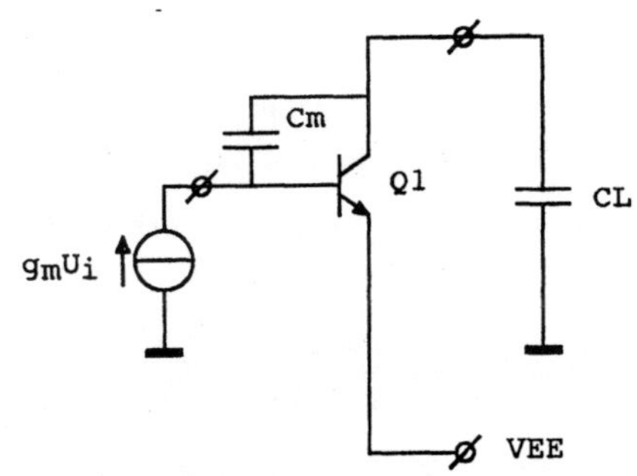

Fig. 3.2. Common-emitter output stage (n-p-n side).

transistor Q_1. The output stage is loaded with capacitor C_L and frequency compensated with Miller capacitor C_m.

The minimum supply voltage of this output stage is about 1.0 V, depending on the size of the emitter area of Q_1 and on the maximum current that has to flow through it. Since the output stage consists of Q_1 only, it is of course very efficient; all current except the base current flows through the output node. The current gain is equal to the current gain β_1 of transistor Q_1. For n-p-n transistors this is in the order of 100, so the input current of an output stage that pulls current of 10 mA

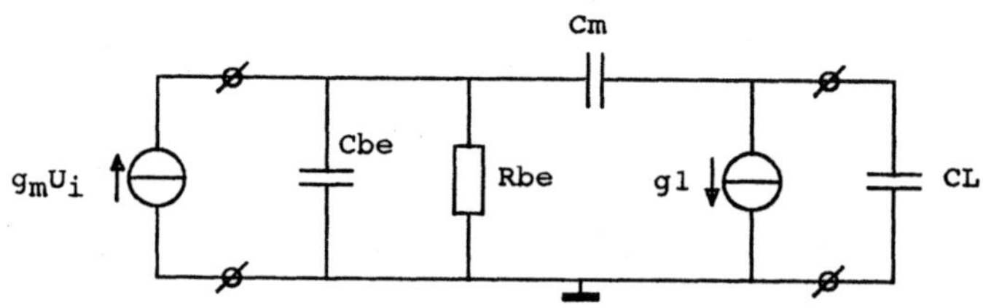

Fig. 3.3. The small-signal equivalent circuit of the common-emitter output stage, using the hybrid-π model.

is 100 μA, which normally will be too much for the intermediate stage. In the case of a lateral p-n-p output transistor, the demands on the intermediate stage are even greater. An output stage comprising just one CE transistor has therefore insufficient gain. Because the CE stage is used in all other output stages discussed in this Chapter, an analysis of the high-frequency behavior will be given.

Fig. 3.3 shows the small-signal equivalent circuit of Fig. 3.2, using the hybrid-π model. The output resistance, as well as other elements to simulate second-order and parasitic effects are not inserted in order to keep the analyses as simple as possible. How the transconductance g_m of the intermediate stage should be determined is explained in Chapter 5; for now it is sufficient that g_m is chosen just small enough to ensure that the gain is smaller than one at the frequency where the phase shift exceeds 180°. When voltage-driven, the intermediate stage is a high-frequency differential pair that does not add significant poles to the transfer. The voltage gain is then given by:

$$\frac{U_o}{U_i} = -\frac{g_m}{sC_m}\frac{1 - s\frac{C_m}{g_l}}{\left(1 + s\frac{C_L}{g_l}\left(1 + \frac{C_{be}}{C_m}\right)\right)} \tag{3.1}$$

If the g_m is chosen correctly, the unity-gain bandwidth follows directly from (3.1):

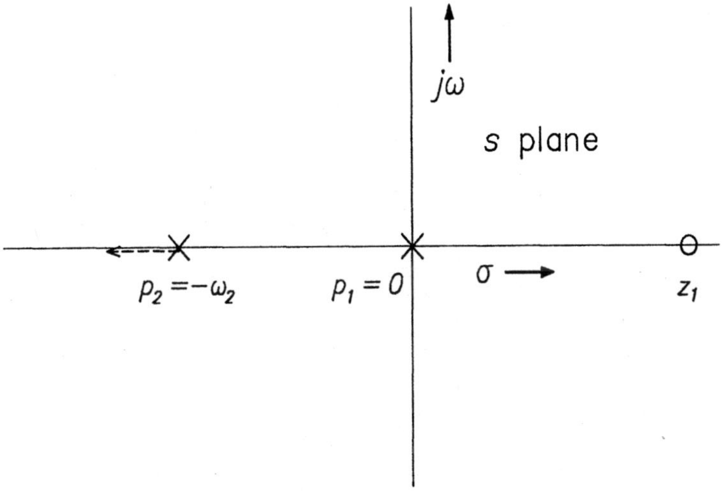

Fig. 3.4. Pole positions of the common-emitter output stage in the s plane (not to scale). The dashed line shows the movement of ω_2 if the transconductance of the output stage increases.

$$\omega_0 = \frac{g_m}{C_m} \tag{3.2}$$

When we write:

$$\omega_2 = \frac{g_o}{C_L\left(1 + \frac{C_{be}}{C_m}\right)} = \frac{g_1}{C_L\left(1 + \frac{C_{be}}{C_m}\right)} \tag{3.3}$$

where g_o stands for the transconductance of the output stage, ω_2 gives the value of the second pole p_2. If, further, the zero in the numerator in (3.1) is neglected because it lies at a frequency much higher than ω_2, the voltage gain becomes:

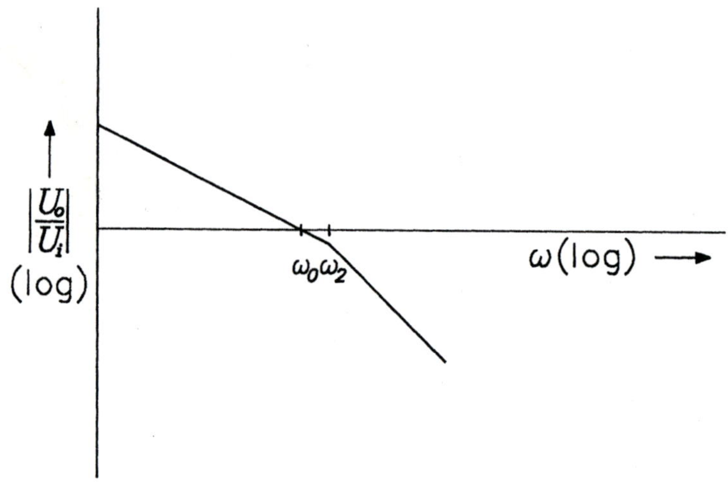

Fig. 3.5. The magnitude of the small-signal voltage gain versus frequency of the common-emitter output stage (not to scale).

$$\frac{U_o}{U_i} = -\frac{\omega_0}{s}\frac{1}{\left(1+\frac{s}{\omega_2}\right)} \tag{3.4}$$

The pole positions in the *s* plane of the CE output stage are shown, not to scale, in Fig. 3.4. Because there is neither an output resistance nor a load resistor assumed in the circuit shown in Fig. 3.2, the low-frequency pole is in the origin. The high-frequency pole is situated at $-\omega_2$ and the high-frequency zero is situated at z_1. The magnitude of (3.4) is depicted in Fig. 3.5. The transfer has a 6 dB/octave frequency roll-off down to ω_2. Because ω_0 always has to be chosen smaller than ω_2 in order to keep a positive phase margin, ω_2 ultimately sets the bandwidth of the output stage, and consequently of the complete OpAmp. This leads to the important conclusion that the bandwidth of any OpAmp with a current-driven output stage is determined by the quiescent current through that output stage and by the maximum load capacitor that has to be driven.

The fact that there are no higher-order poles above the frequency ω_2 is an important advantage of this simple output stage, as will become evident in the following Sections.

Especially in the case of lateral p-n-p output transistors, C_{be} can become large and therefore cause difficulties. This may be illustrated with an example. Suppose transistor Q_1 is a p-n-p transistor with a transition frequency f_T of 50 MHz, which is in fact very good for a lateral p-n-p. If the quiescent current through Q_1 is 100 μA, the transconductance g_o is 3.8 mmho and C_{be} is 12.5 pF. If furthermore, the output stage has to drive a capacitive load of 100 pF and the Miller capacitor C_m is 5 pF, ω_2, given by (3.3), is 1.8 MHz. If we want to increase the bandwidth of the output stage by raising the quiescent current to 1 mA, the high-frequency pole ω_2 is still only 2.4 MHz. This is because C_{be} is now 125 pF and much larger than C_m, thus suppressing effective pole splitting. This can be avoided by choosing a larger value for C_m, but according to (3.2), g_i should then also be increased, in order to keep the same unity-gain bandwidth. The current consumption in the intermediate stage is then also forced to a higher value.

The discussion of the CE output stage showed that its current gain is too low to allow general application. The high-frequency behavior of the output stage is good; there are no limiting high-frequency poles and the bandwidth is solely determined by the transconductance of the output stage and the capacitive load it has to drive. The increase of the C_{be} of p-n-p transistors at high current levels, however, may give rise to problems with the pole splitting.

3.2 DARLINGTON OUTPUT STAGE

To increase the current gain of one single CE stage, it can be preceded by an emitter follower. The Darlington output stage that then emerges is shown in Fig. 3.6. Transistors Q_1 and Q_2 are the CE stage and emitter follower respectively, and the pole splitting is again realized with C_m. Current source I_2 ensures a minimum current through Q_2 at low output currents. As in the previous Section,

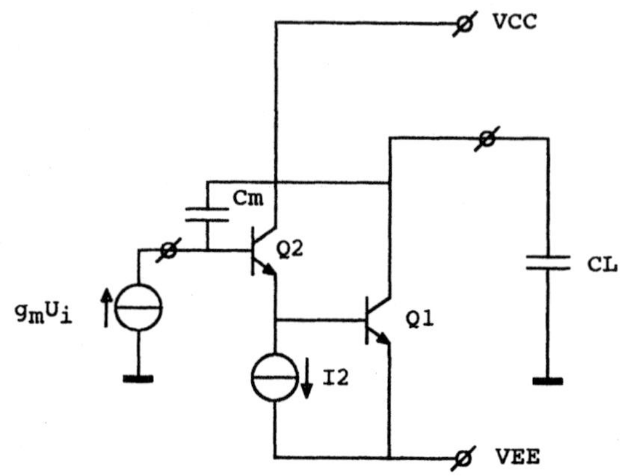

Fig. 3.6. Darlington output stage (n-p-n side).

$g_m U_i$ represents the output of the preceding intermediate stage. The minimum supply voltage for this output stage is about 1.7 V, one diode voltage higher than for the single CE output stage. The current gain is now equal to the product of the current gain of Q_1 and Q_2, $\beta_1\beta_2$. Even if both transistors are lateral p-n-p's with, for instance, a modest current gain of 20, an output current of 10 mA requires an input current of the output stage of 25 μA, which is considerable, but not necessarily too large. The Darlington output stage is economical with the current it pulls from the supply sources; of larger output currents only about $1/\beta_1$ is lost in Q_2. At small output currents, this is relatively more because I_2 is added to that but I_2 is normally not more than about 1/5 of the quiescent current through Q_1.

Fig. 3.7 shows the simplified small-signal equivalent circuit depicted in Fig. 3.6. The value of C_{be2} is assumed small compared to C_{be1}. The transconductance of the output stage g_o is:

$$g_o = \frac{1}{r_{e1} + \frac{r_{e2}}{\beta_1}} \tag{3.5}$$

with $r_{e1,2}$ are the emitter resistances, equal to $1/g_{1,2}$. The resistance at the base of

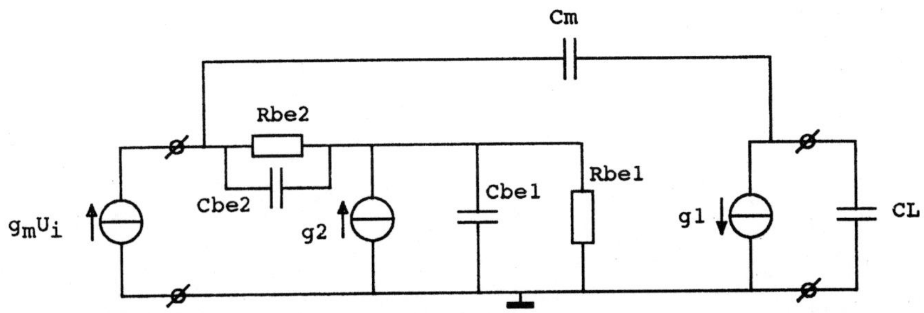

Fig. 3.7. The small-signal equivalent circuit of the Darlington output stage.

Q_1 is defined as:

$$r_p = r_{be1} // r_{e2} = \frac{r_{be1} r_{e2}}{r_{be1} + r_{e2}} \tag{3.6}$$

where the two slashes stand for the two resistors in parallel. The voltage gain is then:

$$\frac{U_o}{U_i} = -\frac{g_m}{sC_m} \frac{\left(1 - s\frac{C_m}{g_o}\right)\left(1 + s\frac{C_{be2}}{g_2}\right)}{\left(1 + s\frac{C_L}{g_o}\left(1 + \frac{C_{be2}}{C_m}\right) + s^2\frac{C_L}{g_o}\left(1 + \frac{C_{be2}}{C_m}\right)r_p C_{be1}\right)} \tag{3.7}$$

The unity-gain bandwidth is again ω_0:

$$\omega_0 = \frac{g_m}{C_m} \tag{3.8}$$

and the pole at the output is ω_2:

$$\omega_2 = \frac{g_o}{C_L\left(1 + \frac{C_{be2}}{C_m}\right)} \tag{3.9}$$

Expression (3.9) for the output pole resembles (3.3) in the previous Section, but now the increase of C_{be1} at large output currents no longer makes the pole splitting ineffective, because Q_1 and C_m are separated by Q_2. The increase in C_{be2} is much smaller, and this hardly deteriorates the pole splitting. There is now a third pole, ω_3, positioned at:

$$\omega_3 = \frac{1}{r_p C_{be1}} \tag{3.10}$$

This internal pole of the Darlington can thought to be positioned at the base of Q_1. The zero due to Q_2 is equal to the transit frequency $2\pi f_T$, which may slightly influence the behavior if p-n-p transistors are used, but which can be ignored if the output stage consists of n-p-n transistors:

$$\omega_T = \frac{g_2}{C_{be2}} \tag{3.11}$$

If this zero and the zero in the right half plane are ignored, (3.5) now becomes:

$$\frac{U_o}{U_i} = -\frac{\omega_0}{s}\,\frac{1}{1 + \frac{s}{\omega_2} + \frac{s^2}{\omega_2\omega_3}} \tag{3.12}$$

Fig. 3.8 shows the pole positions of (3.12) and Fig. 3.9 gives the magnitude of the transfer. At low output currents, the two poles are widely separated, and their positions in Fig. 3.8 are given by $-\omega_2$ and $-\omega_3$, respectively. However, if the output current increases, the two poles move towards each other, and if the condition is met that $4\omega_2=\omega_3$, the poles collide at a frequency that is given by $-2\omega_2$. After the collision the poles move away from the σ-axis and become complex. Note that this is not a root-locus in the conventional meaning. A root-locus emerges as the

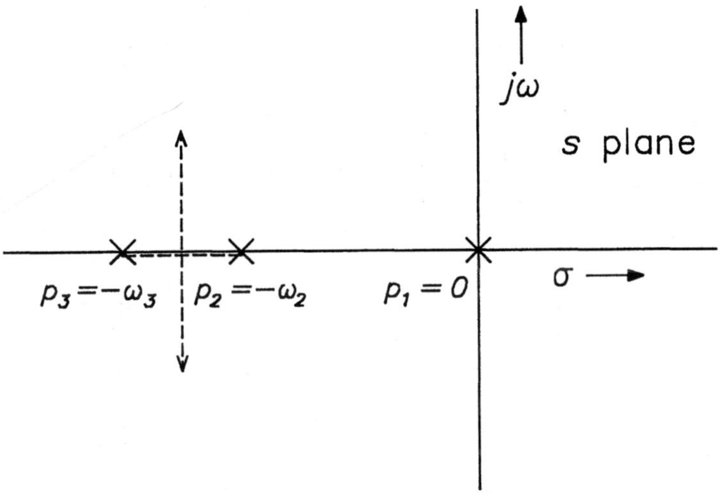

Fig. 3.8. Pole positions of the Darlington output stage in the s plane. The dashed lines show the movement of the poles if the output current of the output stage increases.

low-frequency loop gain of a feedback circuit is varied. The poles of the open-loop transfer are not affected by that. The movement of the poles indicated in Fig. 3.8 arises from the change in the biasing point of the amplifier, and that also alters the positions of the open-loop poles.

The complex poles cause the famous "output bump" [1], as is shown by the dashed line in Fig. 3.9. These complex poles may have a negative influence on the overall performance of the OpAmp. The presence of complex poles causes damped oscillations in the transient response. In practice it is sufficient to keep the damping ratio of the transfer larger than 0.7, which corresponds to an angle of less than 45° in the s plane. This is when the condition is met that $2\omega_2 = \omega_3$. If the current I_2 is small compared to the base current of Q_1, $r_{e2} = r_{be1}$ and if also C_m is large compared to C_{be2}, the output current at which $2\omega_2 = \omega_3$ is:

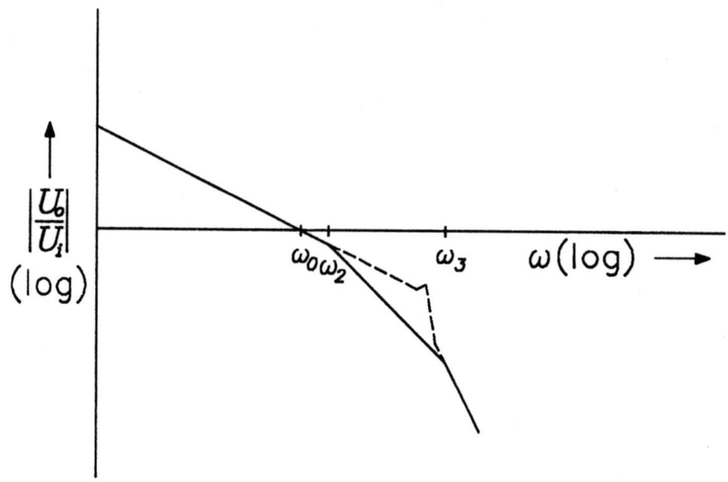

Fig. 3.9. The magnitude of the small-signal voltage gain versus the frequency of the Darlington output stage. The dashed line shows the "output bump" when the output current increases and the poles become complex.

$$I_{bump} = 4\pi \frac{f_T}{\beta_1} V_T C_L \tag{3.13}$$

where V_T is the thermal voltage and f_T is the transition frequency of the transistors. Eq. (3.13) could suggest that if the load capacitor C_L is very small, the poles would become complex at very small output currents; note, however, that the calculations are based on the assumption that C_L is relatively large. For a Darlington consisting of good 5 GHz n-p-n transistors, driving a load capacitor of 100 pF, the value of I_{bump} is only 1.6 mA. When the Darlington consists of lateral p-n-p's, this current is even smaller.

Fig. 3.10 shows a Darlington output stage with an additional Miller capacitor C_{m0} connected between the output and the base of Q_1. In fact, another Miller integrator is created within the Darlington output stage. The load capacitor can therefore thought to be increased, changing (3.9) to:

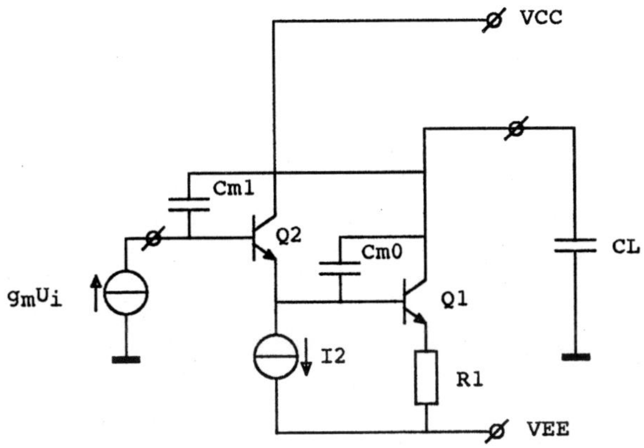

Fig. 3.10. The Darlington output stage with the addition of R_1 and C_{m0}.

$$\omega_2 = \frac{g_o}{\left(C_L + \beta_1 C_{m0}\right)\left(1 + \frac{C_{be2}}{C_m}\right)} \tag{3.14}$$

and (3.10) to:

$$\omega_3 = \frac{1}{r_p\left(C_{be1} + C_{m0}\right)}\left(1 + \frac{\beta_1 C_{m0}}{C_L}\right) \tag{3.15}$$

The output current where the poles become complex, given in (3.13) now is:

$$I_{bump} = 4\pi \frac{f_T}{\beta_1} V_T C_L \left(1 + \frac{\beta_1 C_{m0}}{C_L} + \left(\frac{\beta_1 C_{m0}}{C_L}\right)^2\right) \tag{3.16}$$

The same 5 GHz n-p-n transistors from above together with a C_{m0} of 2.5 pF now yield a maximum output current of 10 mA. A drawback of the insertion of C_{m0} is that it also lowers the second pole at low output currents when this is really not necessary. This means that the bandwidth of the output stage, which should be below the worst-case ω_2, is decreased. In other words, the stability at high output currents has to be paid for with a reduction of the bandwidth.

A second modification of the Darlington that is shown in Fig. 3.10 is the emitter resistor R_1 of Q_1. Insertion of this resistor limits the transconductance of the output stage to $1/R_1$. Thus strong complex poles can be avoided. Resistor R_1 also prevents breakthrough of Q_1 if the collector-emitter voltage becomes large. A disadvantage is that at large output currents the output-voltage range is decreased, so R_1 cannot be chosen too large. Practical values are in the order of 10 Ω.

The Darlington output stage is widely used, and not only in low-voltage applications. The reasons for its popularity have been described above; the DC characteristics are very good. The high-frequency behavior is more troublesome than might have been expected, however. The "output bump" that arises with an increasing output current can be avoided to some extent, but the necessary precautions themselves introduce other problems.

3.3 WIDLAR OUTPUT STAGE

If the supply voltage is less than 1.6 V, the Darlington output stage can no longer be used and a different output stage configuration must be used. The circuit shown in Fig. 3.11 is used in the first low-voltage OpAmp, the LM-10, and we will call it the "Widlar output stage" after its inventor [2], [3]. The output transistor Q_1 has again the CE configuration, to ensure the largest possible output-voltage range. The emitter follower Q_2 is a p-n-p transistor, and its base voltage is very close to the negative supply rail. Transistor Q_3 restores the voltage level so that the output stage can be driven by the intermediate stage. This intermediate stage is modeled by the current source $g_m U_i$, in the way described in the previous Sections. The maximum output current would be reached if all the current of

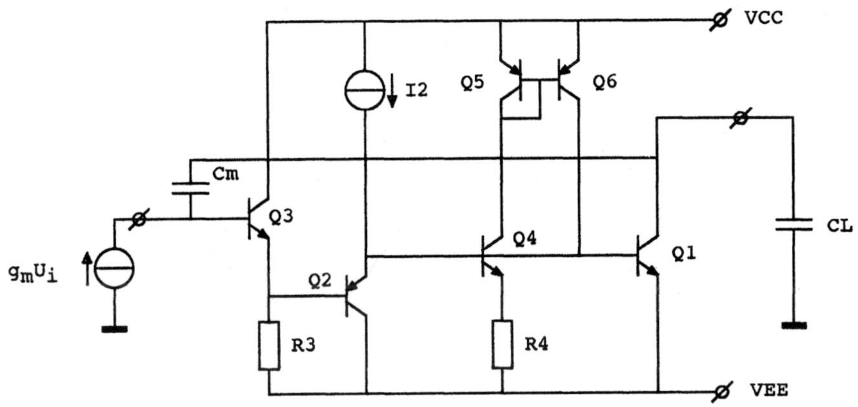

Fig. 3.11. The Widlar output stage (n-p-n side).

current source I_2 were used as base current of Q_1. Therefore a boost circuit, consisting of Q_4, Q_5 and Q_6 is also present. If the output current increases, and a larger base current of Q_1 is needed, the also increasing current through Q_4, Q_5 and Q_6 supplies this base current. The remainder of the boost current is dumped in Q_2. The Miller capacitor C_m splits the input an output poles.

The minimum supply voltage is about 1.0 V, depending on the sizes of the transistors and on the maximum output current that has to be driven. The biasing of the Widlar output stage has to be done experimentally or with the aid of computer simulations. In particular, the fact that the base voltage of Q_2 is dependent on the difference of the base-emitter voltages of an n-p-n and p-n-p transistor makes an analytical estimation of quantities such as the large-signal current gain of the output stage very unreliable. The amount of boost current that is produced by Q_4-Q_6 also strongly influences the biasing of the Widlar output stage, and the non-linear relationships that characterize the behavior of the boost circuit make an analytical interpretation difficult. The current gain of this circuit can be very large. Again, this is strongly dependent on the way the boost circuit is configured. Practical values are in the order of 10^4 to 5.10^4 for an n-p-n output stage and 10^3 to 5.10^3 for p-n-p output stages. The circuit is rather current consuming. The total amount of current lost in Q_2, Q_3, and Q_4-Q_6 can easily be equal to or more than half the current flowing through Q_1.

Before calculating the transfer of this circuit, we focus on the operation of the boost circuit. The positive feedback loop consisting of Q_4, Q_5 and Q_6 might seem to cause instability. The current through Q_4 is mirrored by Q_5 and Q_6 and then flows into the base of Q_1 and Q_4 and into the emitter of Q_2. For large output currents the influence of current source I_2 can be ignored:

$$I_{C4} = I_{B1} + I_{E2} + I_{B4} \approx \frac{I_{C1}}{\beta_1} + I_{C2} \tag{3.17}$$

with I_C the collector current, I_B the base current and I_E the emitter current. But then is also:

$$g_4 = \frac{g_1}{\beta_1} + g_2$$
$$or:$$
$$r_{e4} = r_{be1} // r_{e2} \tag{3.18}$$

where the two slashes stand for the two resistances in parallel. The positive loop gain around Q_4 is then:

$$A\beta_{pos} = \frac{r_{be1}//r_{e2}}{r_{e4} + R_4} = 1 - \frac{1}{1 + \dfrac{r_{be1}//r_{e2}}{R_4}} \tag{3.19}$$

which is always smaller than one. One could also say that the negative resistance on the base of Q_1 and Q_4 introduced by Q_4-Q_6, $r_n=r_{e4}+R_4$ is always an amount R_4 larger than the total positive resistance on that same point determined by Q_1 and Q_2, $r_p=r_{e2}//r_{be1}=r_{e4}$ [4]. The result is that the boost circuit is inherently stable, at least at low frequencies, in spite of the positive feedback. The value of R_4 controls the amount of positive loop gain, and through that also the current gain of the output stage. A small value may be needed, especially in the case of the p-n-p output stage, in order to be able to push large output currents. A small value of R_4 also decreases the high-frequency stability of the circuit, so a compromise must be found. The current source I_2 is also inserted to ensure stability, but now at low output currents. At these low currents R_4 is after all small compared to r_{e4}, and then the loop gain would be equal to one. I_2 prevents this by lowering the value of r_{e2}.

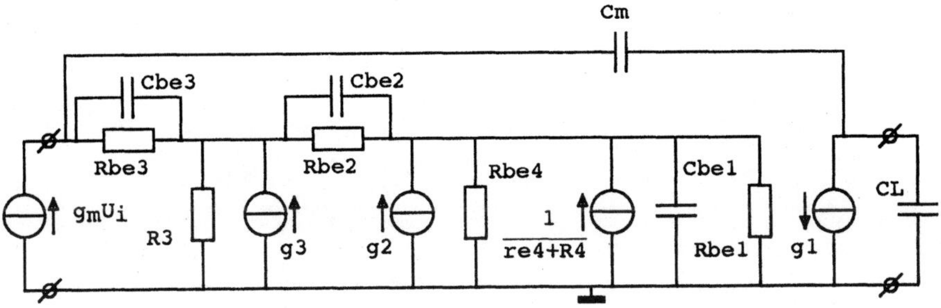

Fig. 3.12. The small-signal equivalent circuit of the Widlar output stage.

Fig. 3.12 shows the small-signal equivalent of the Widlar stage. In the calculations, I_2 is assumed small compared to the current through Q_4, (3.18) is used for g_4 and further C_{be4}, C_{be5} and C_{be6} are neglected because their influence is small compared to the other capacities. The gain is then given by:

$$\frac{U_o}{U_i} = -\frac{\omega_0}{s}\frac{1+\dfrac{s}{\omega_{PNP}}}{1+\dfrac{s}{\omega_2}+\dfrac{s^2}{\omega_2\omega_3}+\dfrac{s^3}{\omega_2\omega_3\omega_4}} \tag{3.20}$$

where as before, ω_0 is the unity-gain bandwidth, given by:

$$\omega_0 = \frac{g_m}{C_m} \tag{3.21}$$

and the output pole ω_2 is:

$$\omega_2 = \frac{g_o}{C_L} \tag{3.22}$$

The transconductance of the output stage g_o is:

$$g_o = \frac{1}{r_{e1} + \dfrac{r_{e2}}{\beta_1}} \cdot \frac{1}{1 + \dfrac{r_{e3}}{R_3}} \cdot \left(1 + \frac{r_p}{R_4}\right) \qquad (3.23)$$

The first factor is equal to the transconductance of a Darlington, as given in (3.6), the second factor renders the decrease caused by emitter follower Q_3 and the third factor shows the influence of the boost circuit. The pole at the base of Q_1 is determined by the total positive and negative resistance at that point and the sum of the base-emitter capacitance of Q_1 and Q_2:

$$\omega_3 = \frac{1}{r_{tot}\left(C_{be1} + C_{be2}\right)} \qquad (3.24)$$

with:

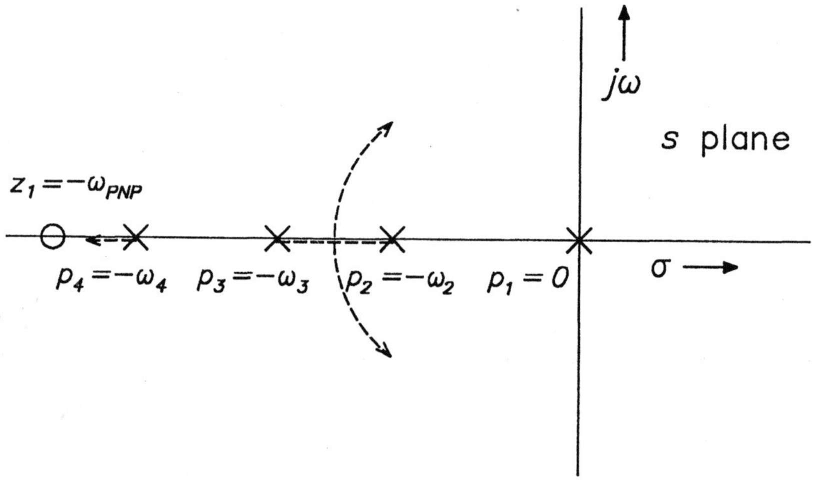

Fig. 3.13. Pole positions of the Widlar output stage in the s plane. The dashed lines show the movement of the poles if the transconductance of the output stage increases.

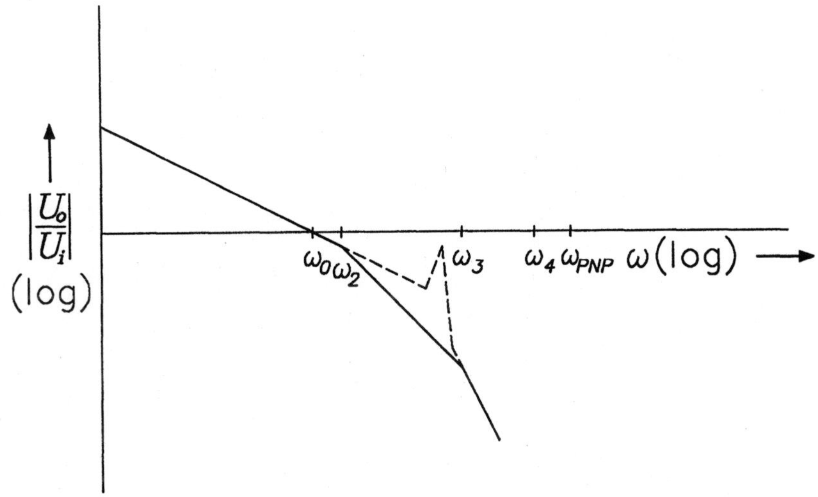

Fig. 3.14. Magnitude of the small-signal voltage gain versus frequency of the Widlar output stage. The dashed line shows the "output bump" when the poles become complex.

$$r_{tot} = r_p // -r_n = \left(\frac{1}{r_{be1} // r_{e2}} - \frac{1}{r_{be1} // r_{e2} - r_{e4}} \right)^{-1}$$
$$= r_{be1} // r_{e2} \cdot \left(1 + \frac{r_{be1} // r_{e2}}{R_4} \right) \tag{3.25}$$

The pole at the base of Q_2 is controlled by the resistance at the base of Q_2 and the base-emitter capacitances of Q_2 and Q_1 in series:

$$\omega_4 = \frac{1}{\left(r_{e3} // R_3\right) \dfrac{C_{be1} C_{be2}}{C_{be1} + C_{be2}}} \tag{3.26}$$

where the two slashes stand for the resistances in parallel. The zero in (3.20) is equal to the transit frequency f_T of a p-n-p transistor; in the case of an n-p-n output stage this zero results from Q_2, and in the case of a p-n-p output stage Q_3 causes the zero:

$$\omega_{PNP} = 2\pi f_T \tag{3.27}$$

If the poles are widely separated, their positions are given by $-\omega_2$, $-\omega_3$ and $-\omega_4$ respectively, as is shown in Fig. 3.13. As the output current, and with that also the transconductance increases, the second- and third-order poles of (3.20) become complex. If $2\omega_2=\omega_3$, the complex poles have an angle of 45° in the s plane. Because the transconductance is increased by the boost circuit Q_4-Q_6, the output current at which this relation is met is much smaller than it was for the Darlington output stage, and the output bump is greater, as is shown in Fig. 3.14. This is again best illustrated by an example. An n-p-n Widlar output stage consisting of 5 GHz n-p-n transistors and 50 MHz p-n-p transistors has to be able to pull an output current of 10 mA at the most. If the current gain of the n-p-n transistors is 100, the boost circuit has to supply a maximum current of 100 μA. Transistor Q_4 has an emitter area that is 1/5 of that of Q_1. Equalizing the base-emitter voltage of Q_4 and the voltage over R_4 to the base-emitter voltage of Q_1 then yields that R_4 should be 750 Ω. If the quiescent current is 300 μA, the current through Q_4 is 27 μA. Because the base current of Q_1 is 3 μA, the current through Q_2 is then 27-3=24 μA. Suppose further that the circuit is dimensioned in such a way that R_3 is 1 kΩ and the current through Q_3 is 50 μA. According to (3.25) and (3.23) r_{tot} is 2 kΩ and g_o is 16 mmho. If the load capacitor is 100 pF, and (3.22) and (3.24) are used to find ω_2 and ω_3, it is found that these are 25 MHz and 23 MHz, respectively. The poles are no longer widely separated, and their positions are therefore not given by ω_2 and ω_3, but have to be calculated using the square law. At an output current of 300 μA, this already gives two strongly complex poles, at 25 MHz, with an angle of 60° in the s-plane.

To prevent the output bump, an additional Miller capacitor can be inserted between the output and the base of Q_1 or between the output and the base of Q_2. This is shown in Fig. 3.15 and the frequency response is shown in Fig. 3.16. It has the same effect as was described in connection to the Darlington output stage. The value of C_{m0} needed at a given output current to prevent the output bump is larger than it was for the Darlington output stage. Analytical evaluation of the minimal

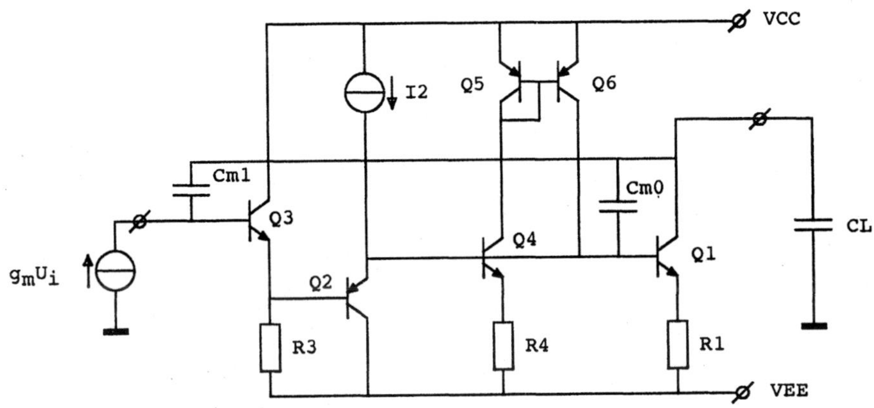

Fig. 3.15. The Widlar output stage with additional Miller capacitor C_{m0} and emitter resistor R_1 (n-p-n side).

value of C_{m0} that prevents complex poles is very cumbersome and not very useful. Determination of C_{m0} with the aid of computer simulations is much more effective. Practical values are in the order of 10 to 20 pF.

The emitter resistor R_1 that was proposed in the previous Section to limit the transconductance to a value of $1/R_1$, can also be inserted in the Widlar output stage, just as was done in the Darlington output stage. As stated there, this resistor does however decrease the output-voltage range of the output stage.

The Widlar output stage is able to operate at a supply voltage of 1 V, a lower value than can be obtained with the Darlington output stage. The current gain of the circuit can be very high, but the current consumption of the complete output stage is much higher when compared to the Darlington output stage. The high-frequency behavior of the Widlar output stage is deteriorated by the two emitter followers that precede the output transistor; this means that the second- and third-order poles need to be split locally with an extra Miller capacitor.

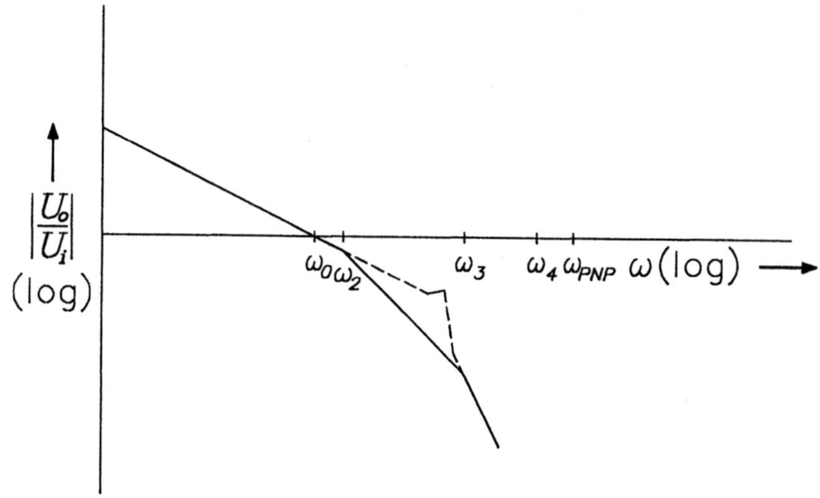

Fig. 3.16. Frequency response of the Widlar output stage with additional Miller capacitor C_{m0}. The dashed line shows the output bump when the poles become complex.

3.4 MULTI-PATH-DRIVEN OUTPUT STAGE

The previous Sections showed that the high-frequency behavior is deteriorated if emitter followers are placed in front of the CE output transistor. This caused problems, especially when the supply voltage is as low as 1 V and two emitter followers had to precede the output transistor. However, the demands on the current gain prohibits the use of just one simple CE stage. The circuit that is discussed in this Section combines the advantages of the simple CE stage with a higher current gain, and therefore bypasses the trade-off between bandwidth and gain.

Fig. 3.17 shows the multi-path-driven (MPD) output stage [5], [6]. The transistors Q_2, Q_3 and Q_4 drive the CE output transistor Q_1 and parallel to this

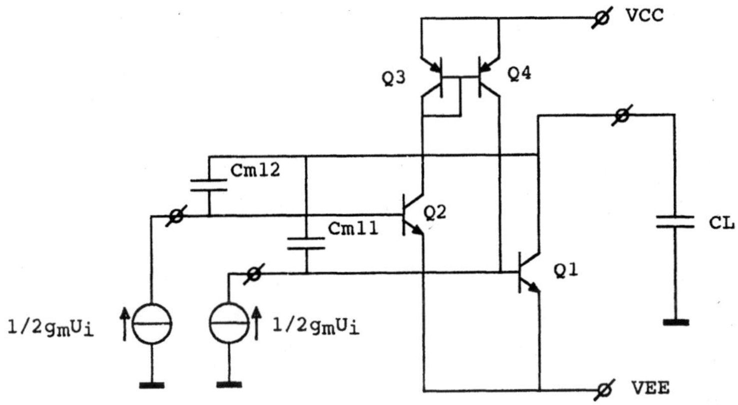

Fig. 3.17. The multi-path-driven output stage (n-p-n side).

path there is a feedforward path directly from the intermediate stage to transistor Q_1. Basically, the path through Q_2-Q_4 supplies the necessary current gain, while the direct path to Q_1 guarantees good high-frequency behavior. The intermediate stage has to supply the output stage with two identical, but uncoupled, input signals. This can for instance be realized with two differential stages in parallel, both driven by the same input voltage, as will be shown in Chapter 4 [7]. The two outputs of the intermediate stage are symbolized in Fig. 3.17 by the two input current sources $1/2g_mU_i$. The poles at the output and at both inputs, are split with Miller capacitors C_{m11} and C_{m12}.

The minimum value of the supply voltage can be as low as 1.0 V, depending of course on the sizes of the transistors and on the maximum output current that has to be driven. The current gain from the bases of Q_1 and Q_2 to the output is $1/2\beta_1(\beta_2+1)\approx 1/2\beta_1\beta_2$, half the value of the Darlington output stage. Only $2/\beta_1$ of the output current is lost in the transistors Q_2-Q_4, which makes the MPD output stage very efficient.

The small-signal equivalent circuit of the MPD output stage is shown in Fig. 3.18. As explained above, there are now two transconductance stages incorporated in the intermediate stage, both of which drive the MPD output stage. For the sake of simplicity, the transistors Q_2, Q_3 and Q_4 are represented by one

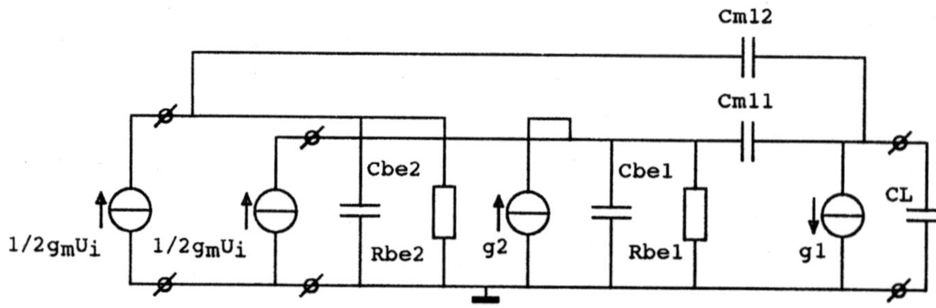

Fig. 3.18. The small-signal equivalent circuit of the multi-path-driven output stage.

compound, non-inverting "transistor" Q_2. At first, an n-p-n MPD output stage is analyzed, and we will further assume that the pole introduced by the current mirror Q_3, Q_4 can be ignored, just as the zero in the right-half plane. The voltage gain is then:

$$\frac{U_o}{U_i} = -\frac{\omega_0}{s} \frac{1 + \frac{s}{\omega_{12}}}{1 + \frac{s}{\omega_{11}} + \frac{s^2}{\omega_{11}\omega_2}} \tag{3.28}$$

The symbols used in (3.28) are, the unity-gain bandwidth:

$$\omega_0 = \frac{g_m}{2\,C_{m12}} \tag{3.29}$$

the output pole:

$$\omega_2 = \frac{g_o}{C_L\left(1 + \frac{C_{be1}}{C_{m11}}\right)} = \frac{g_1}{C_L\left(1 + \frac{C_{be1}}{C_{m11}}\right)} \tag{3.30}$$

the zero:

$$\omega_{12} = \frac{1}{r_{e2}C_{m12}\left(1 + \dfrac{C_{be2}}{C_{m12}}\right)} \tag{3.31}$$

and the pole:

$$\omega_{11} = \frac{1}{r_{e2}C_{m11}\left(1 + \dfrac{C_{be2}}{C_{m12}}\right)} \tag{3.32}$$

If the condition is met that:

$$\begin{gathered} \omega_{11} = \omega_{12} \\ \textit{and:} \\ \omega_{11} \ll \omega_2 \end{gathered} \tag{3.33}$$

the voltage gain (3.28) is equal to the voltage gain (3.4) of the simple CE output stage. The first part of condition (3.33) implies that $C_{m11}=C_{m12}$, which can easily be achieved. If C_{m11} is large compared to C_{be1} and C_{be2}, the second part of (3.33) is met if $\beta_1 C_{m11} \gg C_L$, which was necessary anyway to achieve effective pole splitting. In other words, thanks to the feedforward path from the intermediate stage to Q_1, the high-frequency behavior is not influenced by the presence of Q_2-Q_4. On the other hand, Q_2-Q_4 does enhance the current gain when compared with just the CE stage alone. The trade-off between gain and bandwidth thus seems to be eluded.

The question arises as to whether this result is still valid if the pole at the base of transistors Q_3 and Q_4 of the current mirror is not neglected and if we expand the analysis to the p-n-p MPD output stages. The expression for the voltage gain becomes:

$$\frac{U_o}{U_i} = -\frac{\omega_1}{s}\,\frac{1+\dfrac{s}{\omega_{12}}+2\dfrac{s^2}{\omega_{12}\omega_M}}{1+\dfrac{s}{\omega_{11}}+\dfrac{s^2}{\omega_{11}\omega_2}+2\dfrac{s^3}{\omega_{11}\omega_2\omega_M}} \tag{3.34}$$

The influence of the mirror is displayed by ω_M, the transit frequency of transistors, Q_3 and Q_4, whether they be p-n-p's in the case of an n-p-n output stage, or n-p-n's in the case of a p-n-p output stage. In the latter case, of n-p-n mirror transistors, ω_M is very high, and can normally be neglected.

The second pole can now be approximated by:

$$\omega_2 = \frac{1}{r_{e1}C_L\left(1+\dfrac{C_{be1}}{C_{m11}}\right)+\dfrac{2}{\omega_{PNP}}} \tag{3.35}$$

with ω_{PNP} the transit frequency of the p-n-p transistors. In the case of an n-p-n output stage this factor is introduced by the p-n-p transistors of the mirror Q_3, Q_4 and in the case of a p-n-p output stage it is caused by transistors Q_1 and Q_2. The

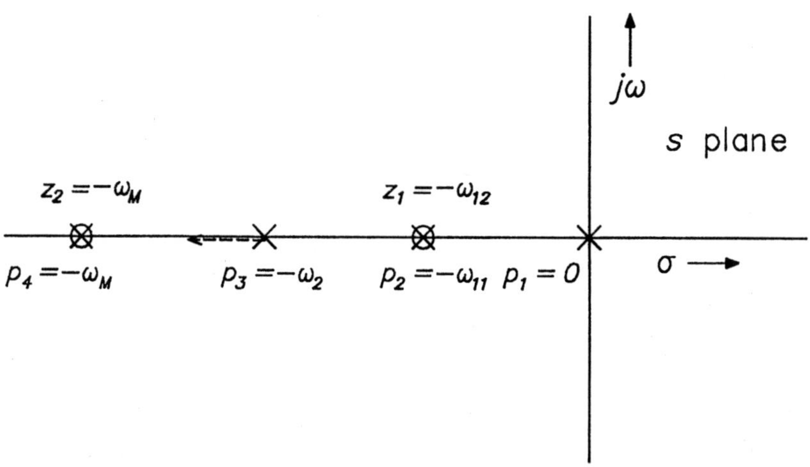

Fig. 3.19. Pole positions of the multi-path-driven output stage in the s-plane.

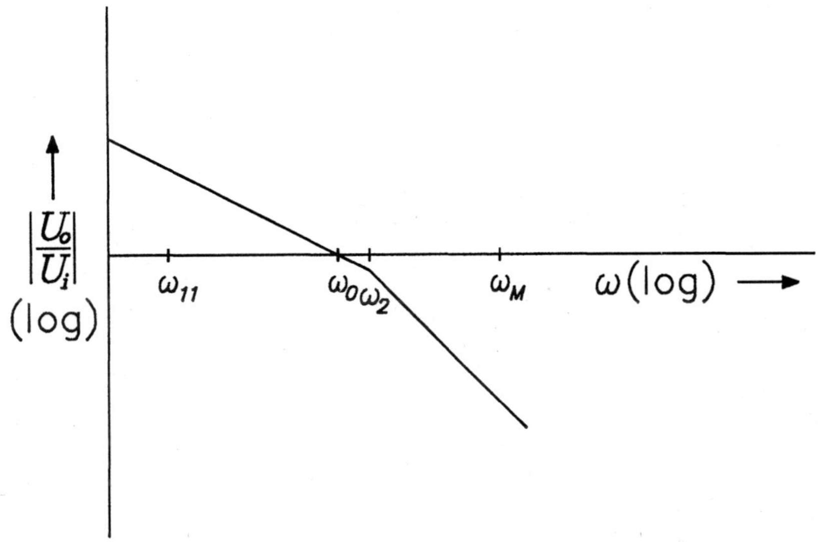

Fig. 3.20. The magnitude of the small-signal voltage gain versus the frequency of the multi-path-driven output stage.

condition that has to be met to cancel the zero ω_{12} and the pole ω_{11} is now:

$$\begin{aligned}&\omega_{11} = \omega_{12}\\ &\textit{and:}\\ &\omega_{12} \ll 1/2\omega_M\\ &\textit{and:}\\ &\omega_{11} \ll \omega_2\end{aligned} \tag{3.36}$$

If (3.36) is satisfied, ω_{11} and ω_{12} cancel, and also the third-order pole and zero cancel one another. This is shown in Fig. 3.19, the pole-zero pattern of the MPD output stage. The increase in the output current, and thus also in the transconductance of the output stage, does not lead to complex poles. This is an important advantage over the Darlington and Widlar output stages. The small-signal voltage gain as function of the frequency, corresponding to (3.34) and (3.36) is shown in Fig. 3.20.

To have this favorable frequency behavior, (3.36) has to be satisfied. The last two of these three conditions impose the strictest demands. Using (3.32) and (3.35) this can then be written as:

$$r_{e2}C_{m11}\left(1+\frac{C_{be2}}{C_{m12}}\right) \gg r_{e1}C_L\left(1+\frac{C_{be1}}{C_{m11}}\right)$$

and:

$$r_{e2}C_{m12}\left(1+\frac{C_{be2}}{C_{m12}}\right) \gg \frac{2}{\omega_{PNP}} \tag{3.37}$$

The implication of (3.37) is best seen if the output current responding with these conditions is calculated. For the n-p-n output stage, the second part of (3.37) limits the output current the most:

$$I_{N\,max} \ll 2\pi f_{PNP} V_T \beta_1 C_{m12} \frac{f_{NPN}}{2f_{NPN} - f_{PNP}} \tag{3.38}$$

Suppose we have n-p-n transistors with a transit frequency f_{NPN}=5 GHz and a current gain β=100, p-n-p transistors with a f_{PNP}=50 MHz and a current gain β=20, and further C_{m11}=10 pF and C_L=100 pF. The output current of the n-p-n MPD output stage then has to stay well below 4 mA. Although this is enough for many applications, output currents as large as 10 mA are sometimes demanded. To allow general application of the MPD output stage, the maximum output current should be increased. This is even more true for the output current of the p-n-p version of the MPD output stage. This current is restricted by the first condition of (3.37). The corresponding output current is now:

$$I_{P\,max} \ll 2\pi f_{PNP} V_T \left(\beta_1 C_{m11} - C_L\right) \frac{C_{m11}}{C_L - C_{m11}} \tag{3.39}$$

With the same transistors as described above, the output current of the p-n-p MPD output stage should be substantially smaller than 90 μA, which is much too small.

The increase of ω_{11} and ω_{12} with the output current can be diminished by inserting a resistor R_2 in series with the emitter of Q_2. This changes (3.32) to:

$$\omega_{11} = \frac{1}{\left(r_{e2}\left(1 + \frac{C_{be2}}{C_{m12}}\right) + R_2\right)C_{m11}} \tag{3.40}$$

The maximum value of the output current of the n-p-n MPD output stage is now:

$$I_{N\,max} < 2\pi f_{PNP} V_T \beta_1 C_{m11} \frac{1}{2 - \frac{f_{PNP}}{f_{NPN}} - 2\pi R_2 C_{m11} f_{PNP}} \tag{3.41}$$

and (3.39) becomes:

$$I_{P\,max} < 2\pi f_{PNP} V_T \left(\beta_1 C_{m11} - C_L\right) \frac{1}{\frac{C_L}{C_{m11}} - 1 - 2\pi R_2 C_{m11} f_{PNP}} \tag{3.42}$$

With emitter resistors R_2=630 Ω or more for the n-p-n output stage and R_2=3 kΩ or more for the p-n-p output stage, the conditions of (3.37) are always met, and the pole ω_{11} and zero ω_{12} cancel. The emitter resistor R_2 allows the cancelation of ω_{11} and ω_{12}, but they limit the minimum value of the supply voltage. At an output current of 1.7 mA, for instance, the voltage over R_2 of an p-n-p MPD stage is 250 mV, which is about the maximum that can be allowed at a 1-V supply voltage.

As can be seen from (3.41) and (3.42), the maximum allowable output current can also be increased by choosing a larger value for C_{m11} and C_{m12}. This does however lower the unity-gain bandwidth as given in (3.29). To maintain the same bandwidth, the transconductance, and with that the current consumption, of the intermediate stage has to be raised. The most effective measure has to be determined depending on the demands on bandwidth, maximum output current and the current consumption of the OpAmp.

The MPD output stage is able to operate at a supply voltage of 1 V, it has a current gain in the same order as the Darlington output stage and it consumes little extra supply current. Further, the circuit has the high-frequency behavior of the CE output stage described in Section 3.1. The conditions that have to be met to ensure that the circuit indeed does not have higher-order poles, can somewhat limit the minimum supply voltage or the maximum output current of the output stage.

3.5 PARALLEL-FEEDBACK CURRENT COMPENSATION

In the first Sections of this Chapter it we demonstrate that in order to improve the gain of the CE stage, the output transistor of the Darlington and Widlar output stages must be preceded by a buffer amplifier. The high-frequency behavior, however, benefits greatly from a direct path from the input to the output transistor. In the previous Section, a feedforward approach is used to combine the direct path to the output transistor with sufficient gain. The circuit shown in Fig. 3.21 makes use of the different strategy. It employs parallel-feedback current compensation (PFCC) [8]. Again, there is a direct path from the intermediate stage to the output transistor Q_1. A second output transistor, Q_2, that also carries a part of the output current is placed parallel to Q_1. The compensation amplifier Q_3-Q_8

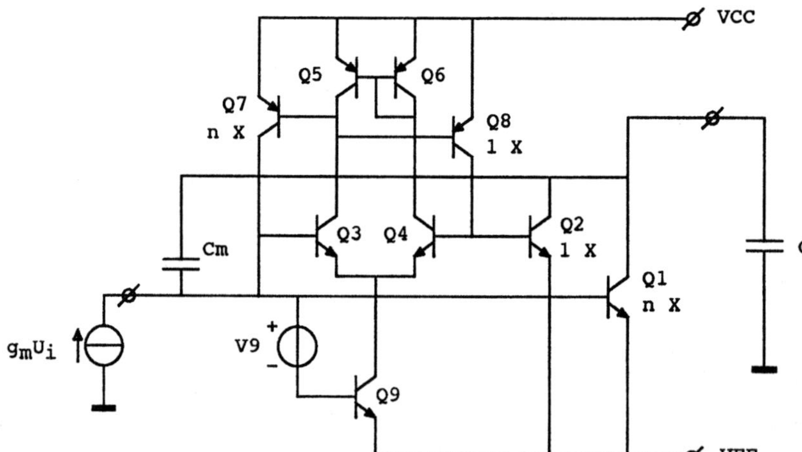

Fig. 3.21. The output stage with parallel-feedback current compensation.

compares the base voltages of the output transistors Q_1 and Q_2, and ensures that their voltage difference is small. One output of the amplifier Q_3-Q_8, the collector of Q_8, supplies Q_2 with the necessary base current. The other output of Q_3-Q_8, the collector of Q_7, is connected to the base of output transistor Q_1. Because the ratio of Q_1 and Q_2 is equal to the ratio of Q_7 and Q_8, 1:n, output transistor Q_1 is then also supplied with sufficient base current. The ratio of Q_1 and Q_2 is chosen relatively large, in the order of 5 to 10, because the direct path from the intermediate stage to Q_1 has to dominate the frequency behavior. The tail-current source Q_9 is proportional to the output current, but much smaller. In Fig. 3.21, this is realized with voltage source V_9 that ensures a constant ratio of the output current I_{out} and the collector current I_{C9} of Q_9. Scaling Q_1 and Q_9 is also possible to realize this constant ratio, but then Q_3 and Q_4 have to be much larger than Q_9, other wise it hat will saturate. As before, the intermediate stage is modeled with the current source $g_m U_i$.

The minimum supply voltage of this output stage is 1 Volt. It is interesting to see that the Early effect caused by an increase of the supply voltage to a value much higher than 1 Volt does not influence the degree of compensation achieved by Q_3-Q_8; all pairs of matching transistors have the same collector voltages. To calculate the current gain of the PFCC output stage, the errors in the compensation of the base current of Q_1 have to be investigated. It can be shown that the unintended current that flows from the collectors of Q_3 and Q_5 into the bases of Q_7 and Q_8 causes the largest error. The base current of Q_7 and Q_8 is $I_{out}/\beta_1\beta_7$ and this current causes a voltage difference between the bases of the differential-pair transistors Q_3 and Q_4. Transistors Q_1 and Q_2 are in parallel with Q_3 and Q_4, and therefore the same voltage difference also causes the collector current of Q_1 to be larger than n times the collector current of Q_2. This induces an input current of the output stage equal to:

$$I_{in} = \frac{I_{out}}{\beta_1\beta_7} \cdot \frac{2}{I_{C9}} \cdot \frac{I_{out}}{\beta_1} \cdot \frac{n+1}{n} \tag{3.43}$$

With m as the ratio of the output current and the current through tail current source Q_9, the total current gain of the PFCC output stage then is:

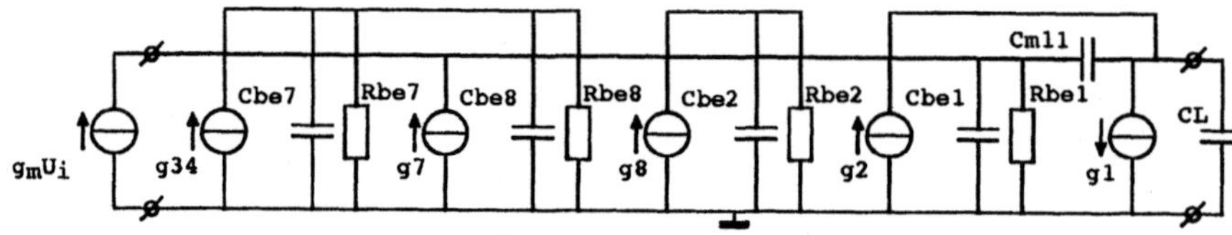

Fig. 3.22. The small-signal equivalent of the output stage with parallel-feedback current compensation.

$$\beta_{tot} = \beta_1 \beta_7 \cdot \frac{\beta_1}{2m} \cdot \frac{n}{n+1} \qquad (3.44)$$

If the β of n-p-n transistors is 100, the β of p-n-p transistors is 20, and m is chosen in the same order as β_1, both the n-p-n and the p-n-p output stage have a current gain of about 10^3. If desired, the compensation amplifier can be extended to diminish the above-mentioned errors and thus the current gain of the PFCC output stage can be increased to 10^4 [9]. Although the output stage seems rather complex, the total amount of current that is drawn from the supplies only exceeds the current through Q_1 and Q_2 by a factor of $2/\beta_1$, which makes the PFCC output stage very efficient.

Fig. 3.22 shows the small-signal equivalent of the PFCC output stage. First, the compensation amplifier Q_3-Q_8 is considered to add no poles and it is modeled by the transconductance stage $g_{3,4}$, g_7, and g_8. The voltage gain is then given by:

$$\frac{U_o}{U_i} = -\frac{\omega_0}{s} \frac{1 + \frac{s}{\omega_{11}}}{1 + \frac{s}{\omega_2} + \frac{s^2}{\omega_2 \omega_3}} \qquad (3.45)$$

with the unity-gain bandwidth:

$$\omega_0 = \frac{g_m}{C_m} \tag{3.46}$$

and the output pole:

$$\omega_2 = \frac{n+1}{n} \cdot \frac{g_1}{C_L} \tag{3.47}$$

and:

$$\omega_3 = \frac{n}{n+1} \cdot \frac{\beta_7}{2m} \cdot \omega_{T1} \tag{3.48}$$

where ω_{T1} is the transit frequency of the output transistor Q_1. Finally, the zero of (3.45) is given by:

$$\omega_{11} = \frac{\beta_7}{2m} \cdot \omega_{T1} \tag{3.49}$$

As could be expected, the output pole ω_2 is not lowered because of C_{be1}; after all, the amplifier Q_3-Q_8 also compensates the current that flows in this diffusion capacity. If m is in the order of β_1 and $n \approx n+1$, the third pole is in the order of the transit frequency of Q_1 and does not cause many problems; especially not because it is at least partially cancelled by the zero ω_{11}.

Fig. 3.23 shows the poles and the zero of (3.45) and the accessory frequency response is depicted in Fig. 3.24.

The analysis leading to (3.45) emanates from a stable compensation amplifier Q_3-Q_8. We will now investigate under which conditions this assumption is correct. The compensation amplifier has a negative feedback loop from Q_4 to Q_6, Q_5, Q_8 and again Q_4, but also a positive feedback loop from Q_3 to Q_7 and again Q_3. At low frequencies, the output impedance of the intermediate stage and the input impedance of Q_9 make sure that the negative loop gain is always larger than the positive loop gain. At higher frequencies the Miller capacitor C_m ensures a stable transfer. But also the phase margin of the negative feedback loop must be positive, otherwise the negative feedback changes into positive feedback. The differential

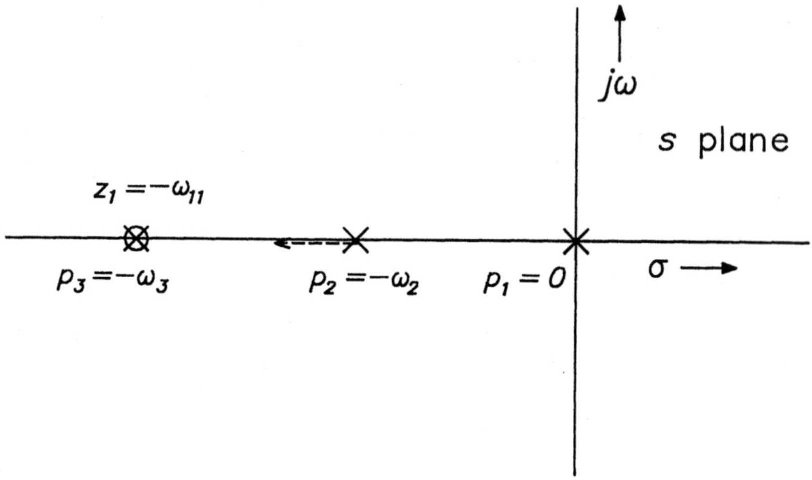

Fig. 3.23. Pole positions in the s-plane of the output stage with parallel-feedback current compensation.

stage Q_3, Q_4 and the mirror Q_5, Q_6 do not add significant poles to the loop gain of the compensation amplifier. The dominant poles are found at the base of Q_8 and at the base of Q_2. It can be seen directly from Figs. 3.21 and 3.22 that these poles have the values ω_{NPN}/β_{NPN} and ω_{PNP}/β_{PNP} both for n-p-n and for p-n-p output stages. The DC loop gain, equal to $\beta_{NPN}\beta_{PNP}/2m$, should not be too large because then the phase margin is negative. This can be realized if:

$$m > 1/2\,\beta_{NPN}^2 \frac{\omega_{PNP}}{\omega_{NPN}} \tag{3.50}$$

For n-p-n transistors with a f_T of 5 GHz and a β of 100, and p-n-p transistors with a f_T of 50 MHz and a β of 20, (3.50) gives that *m* should be larger than 50. If *m* is 100, the current gain of the p-n-p output stage drops to 200, which is rather low. Compensation capacitors can therefore be inserted in the amplifier Q_3-Q_8. This can for instance be done from the base of Q_7 to the base of Q_1 and from the base of Q_8 to the base of Q_2, respectively, although this Miller splitting is not very effective because the voltage gain of Q_7 and Q_8 is only one. Another method to stabilize the compensation amplifier is the insertion of a capacitor from the base

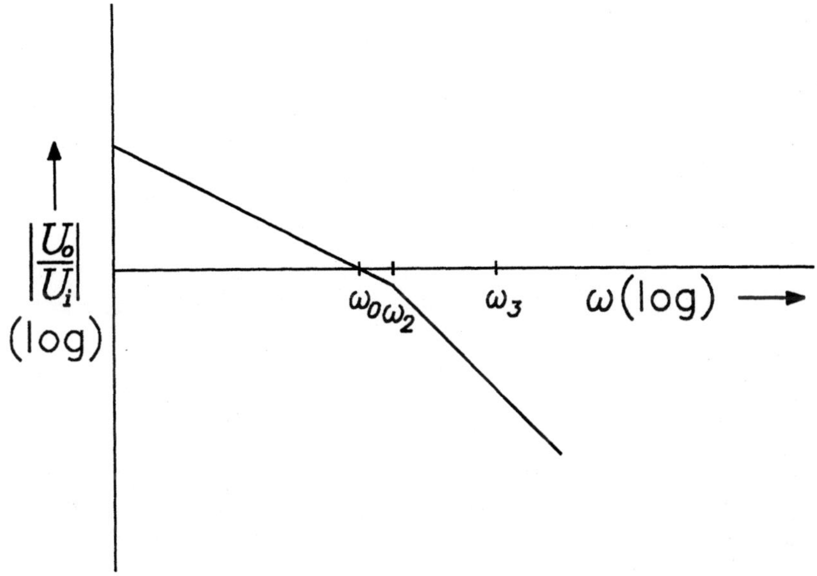

Fig. 3.24. The magnitude of the small-signal voltage gain versus the frequency of the output stage with parallel-feedback current compensation.

of Q_7 and Q_8 to the emitters of Q_3 and Q_4.

The PFCC output stage has potentially a very high current gain, and is relatively efficient for the current it consumes. The high-frequency behavior of the output stage itself is good, but the compensation amplifier, contained in the output stage appears difficult to stabilize.

3.6 CONCLUSIONS

Low-voltage output stages comprise common-emitter output transistors to ensure the maximal output-voltage range. One single CE output transistor does not have the required current gain, and therefore it is preceded by voltage followers in the Darlington and Widlar output stages. The voltage followers add poles to the frequency transfer of the output stage that cause the "output bump" in the frequency response to occur. Two output stages with a direct connection between the intermediate stage and the output transistor have been introduced, of which the multi-path-driven output stage is especially noteworthy because it combines the high-frequency behavior of the CE output stage with the gain of the Darlington, and, further, operates at a supply voltage of only 1 Volt.

3.7 REFERENCES

[1] J. E. Solomon, "The Monolithic Op Amp: A Tutorial Study", *IEEE J. Solid-State Circuits*, Vol. SC-9, pp. 314-332, Dec. 1974.

[2] R. J. Widlar, "Low Voltage Techniques", *IEEE J. Solid-State Circuits*, Vol. SC-13, pp. 838-846, Dec. 1978.

[3] R. J. Widlar, "Low Voltage Compound Invertor Buffer Circuit", U.S. Pat. Appl. Ser. No. 4.228.202, filed October 14, 1980.

[4] J. Fonderie, M.M. Maris, E. J. Schnitger and J. H. Huijsing, "1-V Operational Amplifier with Rail-to-Rail Input and Output Ranges", *IEEE J. Solid-State Circuits,* Vol. SC-24, pp. 1551-1559, Dec. 1989.

[5] J. Fonderie, J. H. Huijsing, "Operational Amplifier with 1-V Rail-to-Rail Multi-path-Driven Output Stage", *IEEE J. Solid-State Circuits*, Vol. SC-26, pp. 1817-1824, Dec. 1991.

[6] J. H. Huijsing and J. Fonderie, "Multi Stage Amplifier with Capacitive Nesting and Multi-Path-Driven Forward Feeding for Frequency Compensation", U.S. Pat. Appl. Ser. No. 654.855, filed February 11, 1991.

[7] J. H. Huijsing, J. Fonderie, "Current Compensation Circuit", Eur. Pat. Appl. Ser. No. 9000351, filed February 14, 1990.

[8] J. H. Huijsing and J. Fonderie, "Current Compensation Circuit", Eur. Pat. Appl. Ser. No. 9000352, filed February 14, 1990.

CHAPTER 4
OTHER CIRCUIT PARTS

Now that the input and output stages have been discussed, the most important circuit parts of the OpAmp are dealt with. What remains is the high-frequency design, which will be examined in the next Chapter, and various circuit parts that are examined in this Chapter. Although these circuits are dealt with in less detail, this does not imply that they are insignificant; it is rather that either the solutions found in literature could be adapted quite simply to low-voltage application, or that the realization of the original circuit appears to be very straight forward, obviating the need for a more detailed explanation.

In the first Section, the purpose and implementation of the intermediate stage is discussed. Several ways to achieve class-AB behavior of the currents through the output-stage transistors are surveyed in the second Section, and the third Section shows how to prevent deep saturation of the output transistors. To prevent the OpAmp from destroying itself if large output currents are loaded or sunk, a circuit to limit this output current can be inserted. Two of these circuits are introduced in the fourth Section. Finally, in the fifth Section, a proportional-to-absolute-temperature (PTAT) biasing circuit is examined.

4.1 INTERMEDIATE STAGES

A simple OpAmp can consist of an input stage, directly coupled to an output stage. Although the simplicity of this topology is attractive, a major disadvantage is that the DC open-loop gain is relatively low, certainly not more than 80 dB. If a larger gain is desired, an additional amplifier stage should be inserted between

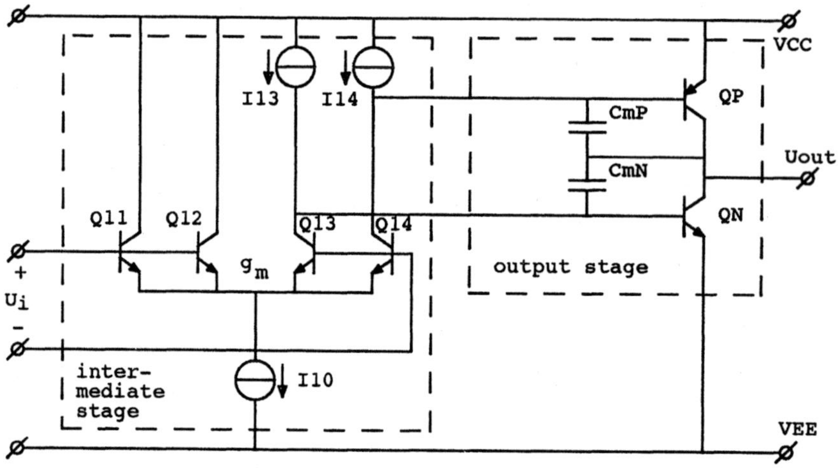

Fig. 4.1. Intermediate stage, followed by a complementary output stage.

the input and output stage. A problem with OpAmps consisting of more than two amplifier stages has always been that the frequency compensation is rather complicated. As will be shown in the next Chapter, a frequency-compensation technique that effectively deals with these problems was developed some years ago, so now the intermediate stage can be inserted without difficulty. One constraint of this frequency-compensation technique is, however, that the intermediate stage should have an inverting input relative to the output voltage. Since the output stage is essentially an inverting CE stage, this implies that the intermediate stage itself should at least have a non-inverting input. This, unfortunately, rules out the use of a simple CE stage. Instead, the intermediate stage should comprise a differential CE stage, with two inputs, and one or two outputs.

Fig. 4.1 shows a differential intermediate stage, together with a complementary output stage. Because both output transistors have to be supplied with decoupled input signals of the same polarity, the intermediate stage consists of two differential stages in parallel. The minimum supply voltage for this stage is determined by the sum of the base-emitter voltage of the p-n-p output transistor, the minimum collector-emitter voltage, or saturation voltage of Q_{13}, and the saturation voltage of the current source I_{10}. The saturation voltage of transistors biased at relatively low currents is in the order of 100 mV, and the base-emitter voltage of the p-n-p output transistor is, depending on the output current it has to push, in the order of 600-800 mV. This yields a minimum supply voltage of only 1 V. The common-mode input voltage of the intermediate stage can for instance be fixed with a control loop that, apart from the intermediate stage itself, also incorporates the summing circuit of the input stage. An example will be given in Chapter 6 when the realized implementations are discussed.

In relation to the low-frequency small-signal behavior of the intermediate stage, its function is to amplify the output current of the input stage, and transfer this to an input current for the output stage. The intermediate stage is therefore basically a current amplifier. As will be shown in the next Chapter, the transconductance of the intermediate stage is also an important parameter, with respect to the frequency compensation of the OpAmp.

The current gain of the intermediate stage is equal to the current gain β_m of the transistors it comprises. The input impedance of the intermediate stage shown in Fig. 4.1 is $2\beta_m r_{e11\text{-}14}$. With a current gain β_m of 100 and a current of 10 μA flowing through each of the transistors Q_{11}-Q_{14}, the input impedance is 500 kΩ. Because this 500 kΩ is small compared to the output impedance of the summing circuit of the input stage, which can be as high as 100 MΩ, a good adaption to the current-source behavior of the output of the summing circuit is guaranteed. The output impedance of the intermediate stage of Fig. 4.1, $r_{o13,14}$, is about 2.5 MΩ, if the Early voltage of the n-p-n transistors is 25 V, and the current through each of Q_{11}-Q_{14} is again 10 μA. If the influence of the current sources I_{13} and I_{14} is taken into account, and those transistors also have an Early voltage of 25 V, the output impedance is halved. Since the input resistance of, for instance, an n-p-n Darlington output stage, biased at a quiescent current of 500 μA, is also as high as 1 MΩ, the adaption to the output of the intermediate stage is not as good, and a loss in the order of up to a factor of two might occur. The total gain of the intermediate stage given in Fig. 4.1 is then in the order of 34 dB.

The transconductance g_m of the intermediate stage shown in Fig. 4.1 is equal to $1/2r_{e11\text{-}14}$, if the current through all branches of the stage is about equal. With a current of 10 μA flowing through each branch, g_m is 0.2 mmho.

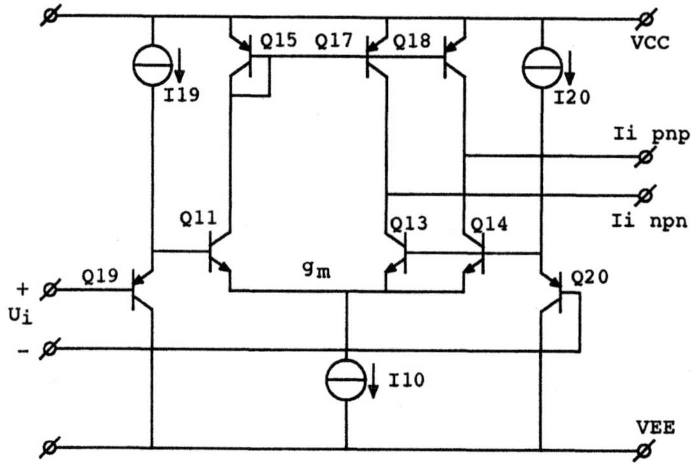

Fig. 4.2. Intermediate stage with current mirrors, consisting of three instead of four branches, and with additional emitter followers.

The transconductance of the intermediate stage can be doubled if it is supplied with current mirrors, as is shown in Fig. 4.2. Another modification of this circuit is that it consists of three, instead of four branches. If the circuit is balanced, and the current through each of the branches is about equal, the transconductance is now $1/r_{e11\text{-}14}$. However, the current consumption is reduced, and the number of transistors used remains the same. A third alteration to the circuit shown in Fig. 4.2 is the addition of the emitter followers Q_{19} and Q_{20}. Although these transistors increase the input impedance of the intermediate stage, and thus deteriorate the adaption at the input, they boost the total current gain of the intermediate stage, and therefore of the complete OpAmp. If the current gain of Q_{19} and Q_{20} is 20, the input impedance of the intermediate stage increases to 10 MΩ, which is still small compared to the 100 MΩ output impedance of the summing circuit. The gain therefore increases by 26 dB.

If the supply voltage of the OpAmp is much higher than the minimum value, the voltages on the matching parts of the intermediate stage diverge. The collector-emitter voltages of Q_{11} and Q_{14}, both coupled to the positive supply voltage, stay equal. The collector-emitter voltage of Q_{13} is, through the n-p-n output transistor,

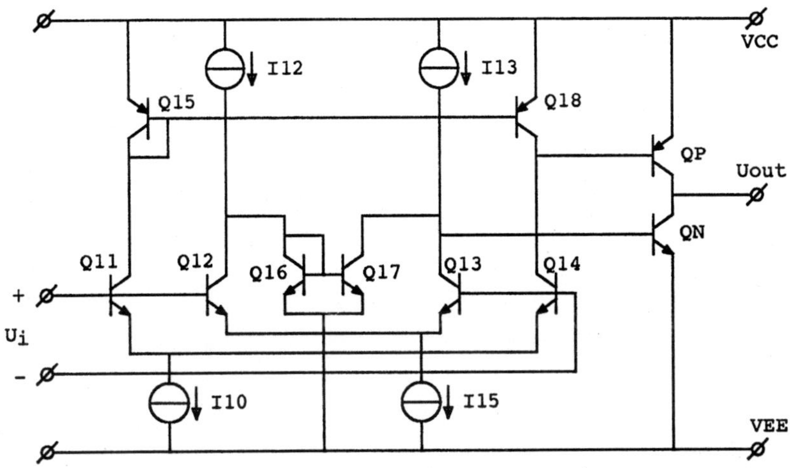

Fig. 4.3. Intermediate stage with separate current mirrors and tail-current sources, to compensate the Early effect on the intermediate-stage transistors.

related to the negative supply rail, however, and at higher supply voltages it is much larger than the collector-emitter voltage of Q_{11}. Because of the Early effect, this difference causes several inconveniences. The most serious is the influence on the class-AB behavior, which will be described in the next Section. The offset-output current that results from the Early effect causes the quiescent current through the output transistors to increase with the supply voltage. To compensate the Early effect, the intermediate stage can be extended by the addition of folded cascodes, much in the same way as this is done in the input stage. The p-n-p cascodes do, however, deteriorate the frequency performance of the intermediate stage. A more simple solution to the problem is shown in Fig. 4.3. The intermediate stage shown in Fig. 4.3 consists of two parts. Transistors Q_{11}, Q_{14} supply the p-n-p output stage, and have current mirrors Q_{15}, Q_{18} that are connected to the positive supply rail, just as the p-n-p output transistor is. The other differential stage, Q_{12}, Q_{13} is connected to the n-p-n output stage and has n-p-n current mirrors Q_{16}, Q_{17}, that are referenced to the negative supply rail, just as the n-p-n output transistor is. In order to be able to fully drive the differential stage Q_{12}, Q_{13}, the current sources I_{12} and I_{13} should be equal to the tail-current sources I_{10} and I_{15}. Although this may seem an unnecessary dissipation of supply power, the consumption is less than when the intermediate stage is equipped with folded cascodes.

4.2 CLASS-AB CURRENT CONTROL-CIRCUITS

As has already been resolved in Chapter 3, a low-voltage push-pull output stage can be realized with complementary CE output transistors. A class-A biasing of this output stage is not permitted, because it would have a quiescent-power consumption that is too high. In a class-B biased output stage, the non-active output transistor is completely shut off, thereby driving the transistors of the intermediate stage into saturation. If this output transistor is then activated again by the reversed input signal, the transient response of the output stage suffers from crossover distortion. The compromise solution is, not surprisingly, the class-AB current control. In a class-AB biased, complementary CE output stage, neither one of the output transistors is ever completely turned off, and still the quiescent power consumption is limited. This Section covers both the conventional feedforward class-AB control and the feedback approach, where the current through the output transistors is measured and, if necessary, adjusted.

4.2.1 Feedforward Class-AB Current Control

Fig. 4.4 shows the general technique for the class-AB biasing of a low-voltage complementary CE output stage [1]. If the resistors R_{12} and R_{22} are assumed small at first, output transistor Q_{11} forms a current mirror together with Q_{12} and the quiescent current through Q_{11} is equal to I_{12}, multiplied by the scaling factor of Q_{11} and Q_{12}. Similarly, the quiescent current through Q_{21} is determined by the current I_{22} and the scaling factor of Q_{21} and Q_{22}. The current sources I_{12} and I_{22} should at least have the same value as the current that can be pushed into, or

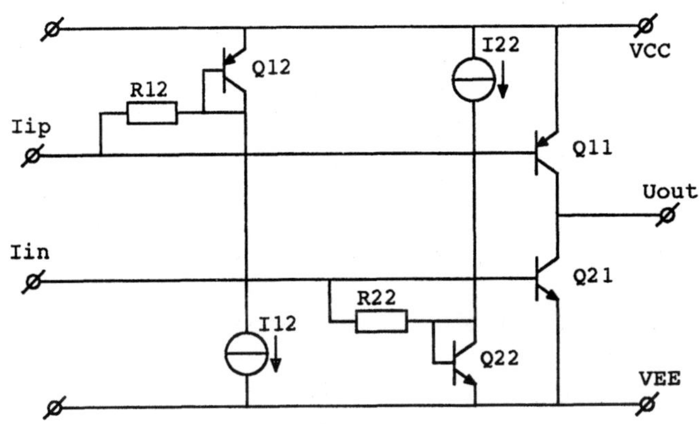

Fig. 4.4. General low-voltage feedforward-biased output stage.

pulled from the output stage by the intermediate stage. In this way, there is always a residual current flowing through the diodes Q_{12} or Q_{22}, and neither one of the output transistors can be completely turned off, even if the affecting output half is deactivated. This typical class-AB current relation is shown in Fig. 4.5.

Naturally, the quiescent current through both output transistors should be equal. Any unintentional mismatch in the circuit parts determining the quiescent current causes the output current to have a non-zero value, when the intermediate stage is not driving the output stage. This offset-output current is compensated by the overall feedback of the OpAmp, and is returned as a (small) offset input voltage. Because the diodes Q_{12} and Q_{22} are matched with the output transistors Q_{11} and Q_{21}, respectively, temperature variations do not alter the biasing of the output transistors. A change in the supply voltage could influence the value of the current sources I_{12} and I_{22}, which in return changes the quiescent current of the output stage. The current sources should therefore be designed to be relatively insensitive to these supply-voltage changes.

The diodes Q_{12} and Q_{22} shunt the input impedance of the output stage and therefore lower the gain. If the class-AB transistors Q_{12} and Q_{22} are scaled 1:10 with the output transistors Q_{11} and Q_{21}, the current gain of the output stage is reduced to 10. Resistors R_{12} and R_{22} help to limit this reduction of the gain by

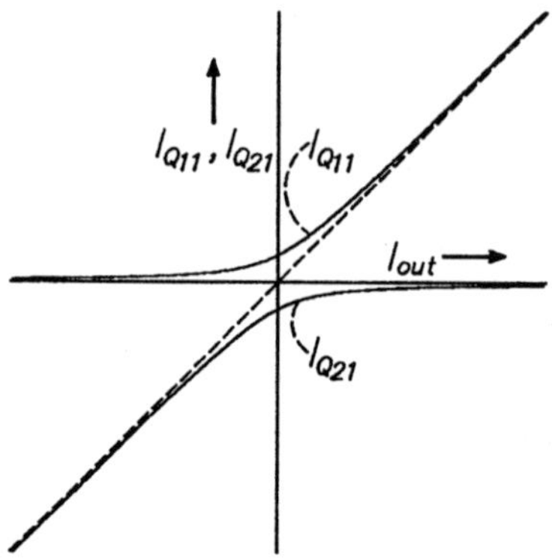

Fig. 4.5. Current relations of the class-AB current regulation. Neither one of the currents through the output transistors is ever completely cut off.

increasing the impedance at the bases of the output transistors. A large resistor value thus benefits the DC gain of the output stage. However, if the intermediate stage pushes or pulls almost all of the current of either one of the current sources I_{12} or I_{22} through resistor R_{12} or R_{22}, respectively, the voltage drop over the resistors should not be so large that either I_{12} or I_{22} is saturated. At a minimum supply voltage of 1 V, for instance, the maximum voltage over the resistors is equal to the supply voltage, minus one base-emitter voltage of Q_{12} or Q_{22} and one saturation voltage of I_{12} or I_{22}, which is about 400 mV. If the intermediate stage can maximally push or pull 10 μA, I_{12} and I_{22} should also be 10 μA, and the maximal value of the resistors is 40 kΩ. Compared to the 2.5 kΩ impedance of the diode Q_{12} or Q_{22} alone, this is a great improvement. The loss in current gain β_o of the output stage is now:

$$\frac{\Delta\beta_o}{\beta_o} = \frac{r_{io}}{r_{io} + R_{12,22}} \tag{4.1}$$

if r_{io} is the input resistance of the output stage, without the class-AB circuitry. If the current gain of the output transistors in Fig. 4.4 is 100 and the quiescent current is 100 μA, the input resistance of the output stage is 2.5 kΩ, so with the above calculated value of R_{12} and R_{22} of 40 kΩ, the current gain is reduced by 38%, or about 4 dB.

The resistors R_{12} and R_{22} also cause a deviation of the intentioned value of the quiescent current, because there is a small difference in the base-emitter voltages of Q_{11} and Q_{12}, and Q_{21} and Q_{22}, respectively. For the n-p-n half of the output stage, this voltage difference is equal to the product of the base current of the output transistor Q_{21} and the resistor this current flows through, that is R_{22}. If the resistors R_{22} is again 40 kΩ, the scaling of Q_{22} and Q_{21} is 1:10, and the current of the current source I_{22} is 10 μA, the quiescent current should ideally be 100 μA. However, with a current gain in the output stage of 100, a voltage over the resistors of 19 mV causes the quiescent current to drop to 47% of 100 μA, or 47 μA. The same applies to the p-n-p output stage, where the deviation of the intentioned value of the quiescent current is likely to be even larger, because the current gain of the p-n-p transistors is lower.

Suppose the single output transistor in the circuit depicted in Fig. 4.4 is replaced by a more complex output stage, such as, for instance, the Widlar output stage. The diodes Q_{12} and Q_{22} now bias the first emitter followers instead of the output transistors of the Widlar output stage. Since the exact current gain of this output stage is difficult to predict, the current through the actual output transistors is also difficult to calculate accurately. Furthermore, the input impedance of this output stage is much higher, and therefore the loss in DC gain caused by the diodes Q_{12} and Q_{22} is now quite considerable.

We can conclude that the feedforward class-AB current control shown in Fig. 4.4, is attractive because of its simplicity, but it is not able to accurately control the current through the output transistors of an output stage more complex than the simple CE stage. Furthermore, the feedforward biasing deteriorates the signal behavior of the output stage.

4.2.2 Feedback Class-AB Current Control

In Section 4.2.1, it was shown that the circuit to bias a low-voltage output stage in a feedforward manner is small and easy to implement. However, especially if a compound output stage is used, its accuracy is limited and it degrades the DC gain. A different approach to achieve the class-AB output-current relation is shown in Fig. 4.6 [1]. A Darlington output stage consists of a p-n-p output transistor Q_{11} and driver Q_{12}, and of an n-p-n output transistor Q_{21} and driver Q_{22}. The base-emitter voltage of the p-n-p output transistor Q_{11} is measured by Q_{33}, and transferred to Q_{34}, where it reappears as the base-emitter voltage of this transistor. The sum of this transferred base-emitter voltage of Q_{11}, and the base-emitter voltage of n-p-n output transistor Q_{21}, drives the base of Q_{32}, where it is compared to the reference voltage V_R, by the differential amplifier Q_{31}, Q_{32}. When the sum of the base-emitter voltages of both output transistors is unequal to V_R, a correction signal is steered to both Darlington pairs by the feedback amplifier Q_{31}, Q_{32}. The sum of the base-emitter voltages of Q_{11} and Q_{21}, and thereby also the product of their collector currents is thus kept constant, and the class-AB current relation is realized. In formula:

$$I_{Q_{11}} \times I_{Q_{21}} = constant \tag{4.2}$$

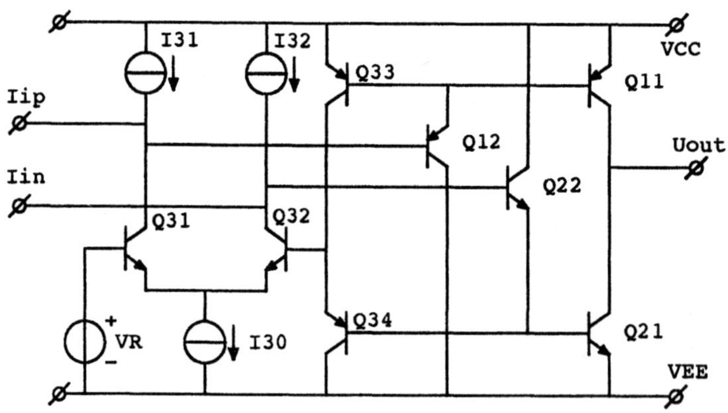

Fig. 4.6. Feedback-biased Darlington output stage.

Fig. 4.5, showing the current relations of the feedforward biased output stage, also applies to the relation defined in (4.2).

The feedback-control loop operates independently of the feedforward driving by the intermediate-stage transistors, or in other words, the functions of biasing and signal driving are separated. The signal driving benefits from this method, because the inputs of the two output pairs are directly accessible, and the signal does not need to pass through circuit elements, which are for biasing purposes only. This is not always true when feedforward biasing is employed. The biasing itself is improved because it is independent of the driving characteristics of the output stage. If the reference voltage V_R is composed of two diodes that are matched to the output transistors Q_{11} and Q_{21}, it is also independent of variations in temperature and process parameters like for instance the current gain β. Because the current through the output transistors is measured directly at their bases, the influence of supply-voltage variations on the performance of the output stage itself, is also largely cancelled. These supply-voltage variations still have their effect on the differential amplifier Q_{31}, Q_{32}, but, if necessary, this can be compensated by using cascodes, as will be shown later on.

The feedback biasing has an additional advantage over the feedforward class-AB control. Suppose the intermediate stage drives the nodes I_{ip} and I_{in} in Fig. 4.6 with two small in-phase signals $+i$. The current in node I_{ip} cannot flow into the driver Q_{12} because this is of the p-n-p type, and it must therefore go through the differential-pair transistor Q_{31}. The opposite signal, that is $-i$, then flows through the other half of the differential pair, Q_{32}. This means that the n-p-n emitter follower Q_{22} is not driven with $+i$, but with $+2i$; the feedback amplifier thus doubles the input signal that is supplied to the output transistors by the intermediate stage.

The minimum supply voltage of the circuit shown in Fig. 4.6 is about 1.8 V, which is not surprising, because the output stage comprises two Darlington pairs and makes use of the summing of two base-emitter voltages. The circuit shown in Fig. 4.7 operates at supply voltages as low as 1 V [2], [3]. For the sake of simplicity, the output stage is composed of just two single CE output transistors, Q_{11} and Q_{21}, but the use of this class-AB current control can be extended to all kinds of output stages.

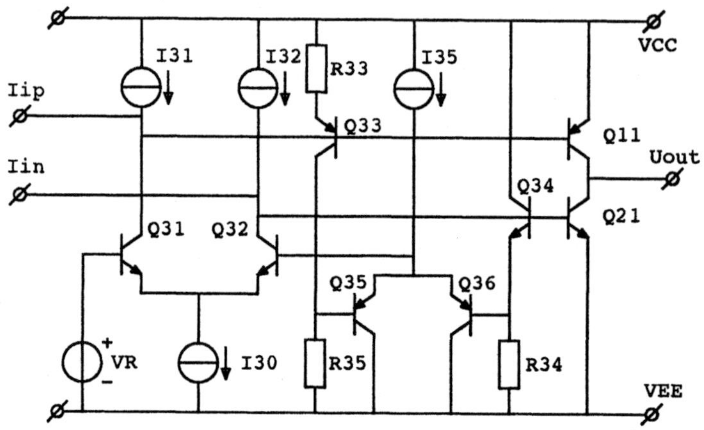

Fig. 4.7. Feedback-biased CE output stage, with residual current that is half the quiescent current.

An image of the current through p-n-p output transistor Q_{11} is transferred by Q_{33} into a voltage over R_{35}, and this voltage is compared with the voltage over R_{34}, which is proportional to the current through the n-p-n output transistor Q_{21}. Suppose n-p-n output transistor Q_{21} is driven hard. The voltage over R_{34} is then higher than the voltage over R_{35}, and all the current from current source I_{35} flows through Q_{35}. The voltage at the emitter of this transistor is made proportional to the reference voltage V_R by the differential feedback amplifier Q_{31}, Q_{32}. The voltage over R_{35} is then fixed, and since it is coupled to the current through the p-n-p output transistor Q_{11}, that too is then proportional to V_R. Similarly, if Q_{11} is driven, the current through Q_{21} is made proportional to V_R. The current through the non-active output transistor is thus dependent on V_R only, and, in particular, it is independent of the current through the active output transistor.

If the current through both output transistors is equal, and consequently the output current is zero, this quiescent current is twice the residual current through an output transistor if the other is driven hard. After all, if both the currents through Q_{11} and Q_{21} are equal, the current I_{35} equally distributes over Q_{35} and Q_{36}. The voltages over the resistors R_{35} and R_{34} are then higher than in the situation described above, and this results in a higher current through the resistors.

This higher current is then transferred in a quiescent current through Q_{11} and Q_{21}, which is also two times higher. To be independent of variations in temperature and of process variables, the reference voltage V_R should be composed of one diode matched to Q_{35} and Q_{36}, and a resistor matched to R_{35} and R_{34}.

A second advantage of the circuit shown in Fig. 4.7 over that shown in Fig. 4.6 is that its high-frequency behavior is easier to control. In Fig. 4.6, the impedance on the base of Q_{32} can become very high, if the n-p-n output transistor Q_{21} is driven hard. This introduces a second dominant pole in the control loop through Q_{32}, Q_{31}, Q_{12} and Q_{33}, that regulates the current through the p-n-p output transistor Q_{11}, and instability might occur. In the circuit depicted in Fig. 4.7, the loop through Q_{32}, Q_{34} and Q_{36}, controlling the residual n-p-n current, and the loop through Q_{32}, Q_{31}, Q_{33} and Q_{35}, controlling the residual p-n-p current, both have only one dominating pole frequency, and, consequently, the circuit has a stable frequency response.

Both circuits shown in Figs. 4.6 and 4.7 suffer from the Early effect on their feedback amplifiers Q_{31}, Q_{32} and consequently, the quiescent current through the output transistors varies with the supply voltage. The circuit given in Fig. 4.8 does not have these drawbacks, because the feedback amplifier is cascoded. The current through the output transistors Q_{11} and Q_{21} is measured the same way as before,

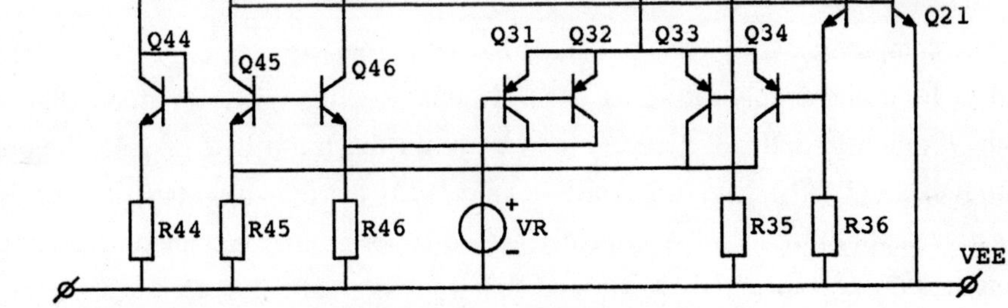

Fig. 4.8. Cascoded feedback-biased CE output stage.

but now the decision pair, selecting the smallest of the output-transistor currents, is incorporated in the differential feedback amplifier Q_{31}-Q_{34}. This class-AB circuit functions the same way as that shown in Fig. 4.7; the residual current through the non-active output transistor is half the quiescent current through both output transistors, and it is determined only by the reference voltage V_R, and not by the current through the active output transistor. The correction signals from the feedback amplifier are supplied to the cascode Q_{45}, Q_{46}, and, through that, to the p-n-p and n-p-n part of the output stage. The intermediate stage, preceding the class-AB current control and the output transistors, is also cascoded by Q_{42}, Q_{43}. All matching parts of the intermediate stage and the class-AB current control now have the same collector voltages, and the Early effect is effectively compensated.

The price that has to be paid for this independence of supply-voltage variations is the extension of the signal path through the common-base stages. This introduces an extra pole in the transfer of the output stage, and in Chapter 3 it was argued that this should preferably be prevented. The combination of the circuits shown in Figs. 4.3 and 4.7 yields an intermediate stage, class-AB current control and an output stage, which is also insensitive to supply-voltage variations, but, nevertheless, with a direct path from the intermediate-stage transistors to the output-stage transistors. This combination can be advantageous over the circuit of Fig. 4.8 and it will be shown in Chapter 6.

4.3 OUTPUT-SATURATION PROTECTION

The complementary CE output stage has an output-voltage range that reaches almost from one supply rail to the other. In this way, the relatively small supply-voltage range is optimally used. Fig. 4.9 shows the n-p-n half of a Darlington output stage Q_1, Q_2, driving a resistive load R_L. If the output voltage is near the negative supply rail, then consequently, the collector-emitter voltage of the output transistor Q_1 becomes very small, and this transistor saturates. Heavy saturation of the output transistor should be avoided, because the accessory charge that is

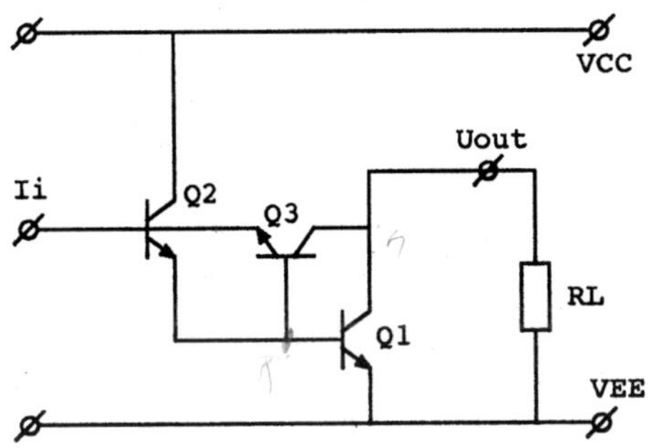

Fig. 4.9. Transistor Q_3 protects the output transistor Q_1 against heavy saturation.

then stored in the base of the transistor takes some time to get carried off, if the input signal of the output stage reverses; this results in a undesirable delay between the input and output of the output stage. Also, the parasitic substrate p-n-p transistor becomes forward biased, causing a large current to be injected into the substrate. This results in a strong reduction of the current gain of the output stage. To prevent heavy saturation, the voltage over the base-collector junction of Q_1 is measured by transistor Q_3. If this base-collector junction tends to get forward biased, the same happens to the base-collector junction of Q_3. This transistor is then biased in the inverse mode; the current through Q_3 flows from the emitter to the collector, removing the input current of the driver transistor Q_2, and, with that, the cause of the saturation. Because both transistors Q_1 and Q_3 are of matching n-p-n types, the control loop is independent of process parameters. The ratio of the emitter areas of Q_1 and Q_3 controls the amount of saturation which is permitted; satisfactory transient behavior is still achieved if Q_3 is ten times smaller than Q_1.

Naturally, this saturation-preventing transistor can also be inserted in output stages other than the Darlington, and both in the n-p-n and in the p-n-p half.

4.4 OUTPUT-CURRENT LIMITERS

If the output stage is driven very hard, the large currents flowing through it, cause a dissipation, which might destroy the OpAmp. To prevent this from happening, the current through the output transistors should be limited. Normally, this is done by inserting a resistor in series with the collector or emitter of the output transistors, and sensing the voltage over this resistor. If the voltage exceeds a threshold value, action is undertaken to limit the input current of the output stage.

This is not possible in the case of a low-voltage output stage, because it would result in a undesirable reduction of the output-voltage range. The sensing of the current through the output transistors has to be done in parallel, on the bases of the transistors. The current through Q_1 of the Darlington output stage shown in Fig. 4.10, is measured by Q_3 [8]. This transistor is of the same type, but smaller than Q_1. If the current through Q_1 increases, the voltage over R_4 eventually allows the current mirror Q_4, Q_5 to conduct. The second current mirror Q_6, Q_7 then removes the input current of the emitter follower Q_2, which drives Q_1, and in this way the current through Q_1 is limited. The threshold, built in by resistor R_4,

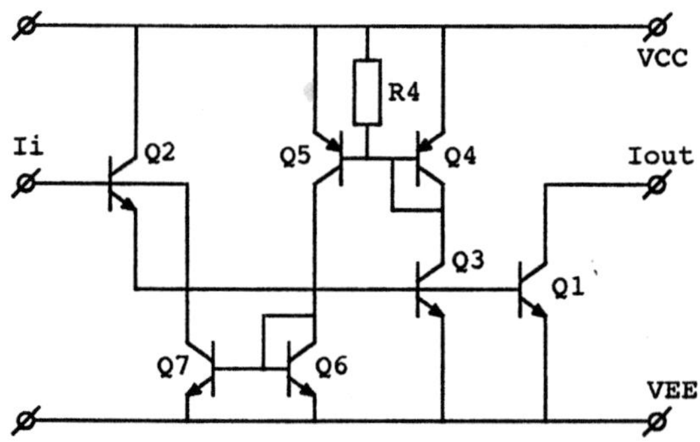

Fig. 4.10. Output-current limitation; the current through Q_1 is sensed with Q_3, and limited by removing the input current of Q_2.

guarantees that the influence of the current-limitation circuit below the maximum current is very small.

Suppose that the maximum allowable current through Q_1 is 20 mA, that Q_3 has an emitter area ten times smaller than Q_1, and that Q_4 and Q_5 start to conduct at an base-emitter voltage of 600 mV. The value for R_4 is then 300 Ω.

Although the current limiter shown in Fig. 4.10 is effective, it uses as much as six transistors, and its high-frequency behavior is not inherently stable, because of the rather long control loop that is created. This generates oscillations in the transient response when the current limiter is activated. A more simple solution is shown in Fig. 4.11. The n-p-n Darlington output stage is again used to demonstrate the functioning of the current limiter, but the use of the circuit can be extended to other output stages, both of the n-p-n and p-n-p type. The current through Q_1 is again measured by the parallel transistor Q_3. If the current through Q_1 and Q_3 increases, Q_3 eventually saturates, because of the voltage drop over R_4. This saturation is sensed by Q_4, in the same way as described in Section 4.3. When the voltage over R_4 is large enough to enable Q_4 to operate in the inverse mode, this transistor removes the input current of Q_2, and therefore prevents a further increase in the current through Q_1. The limiter comprises only two

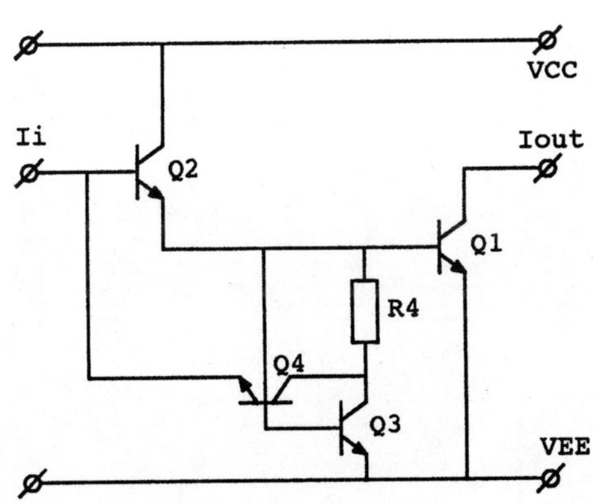

Fig. 4.11. Improved current-limitation circuit. The circuit uses only two transistors, and has a stable transient response.

transistors, and its transient response is much improved: the control loop through Q_4 and Q_2 is much smaller than the loop of the circuit shown in Fig. 4.10, and the phase margin of the loop gain is therefore larger, yielding a transient response without oscillations.

If the output current should, for instance, again be limited to 20 mA, and Q_3 is one-tenth of Q_1, the value of resistor R_4 is again 300 Ω, the ratio of 600 mV and 2 mA.

4.5 BIAS CIRCUIT

The biasing currents of an OpAmp have to be derived from a reference-current source. This reference-current source should be independent of variations in the supply voltage, and preferably also of variations in temperature. Although many other possibilities are available, the proportional-to-absolute-temperature (PTAT) current source is normally used to accomplish this function. The circuit shown in Fig. 4.12 is a PTAT current source, which is insensitive to supply-voltage

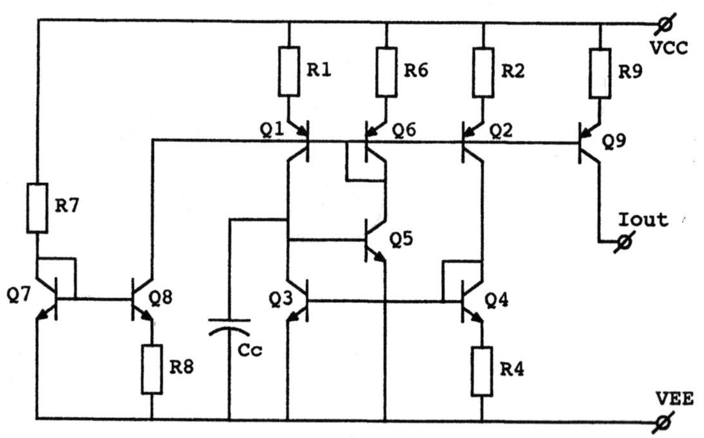

Fig. 4.12. Proportional-to-absolute-temperature current generator.

variations [4]. The feedback loop through Q_5, Q_6, Q_2, Q_4 and Q_3 forces the current through the branch Q_1, Q_3 to be equal to that through the branch Q_2, Q_4. The voltage over R_4 is determined by the difference in the base-emitter voltage of transistors Q_3 and the double transistor Q_4. The current through each of the branches Q_1, Q_3 and Q_2, Q_4 is then:

$$I_{PTAT} = \frac{V_T \ln 2}{R_4} = \frac{kT \ln 2}{qR_4} \tag{4.3}$$

where V_T is the thermal voltage.

The circuit has a positive-feedback loop through Q_5 and Q_1, with a loop gain that has the DC value:

$$A\beta_{pos} = \beta_5 \tag{4.4}$$

if the current gain β of all transistors is assumed large. The loop gain of the negative loop is:

$$A\beta_{neg} = \beta_5\left(1 + \frac{R_4}{r_{e4}}\right) = \beta_5(1 + \ln 2) \tag{4.5}$$

where (4.3) is used to obtain the last equality. This is always larger than the gain of the positive feedback loop. The circuit is thus stable, although a compensation capacitor between the base of Q_5 and ground may be needed to bypass high-frequency instability.

To avoid the state where the currents through all transistors are zero, a start-up circuit Q_7, Q_8 forces a small current through the p-n-p current mirrors. This does not affect normal operation, as long as the start-up current is much smaller than I_{PTAT}.

Variations in the supply voltage have very little influence on the operation of the circuit shown in Fig. 4.12, because the collector voltages of all matching transistors are equal. The minimum supply voltage of the circuit is about 1 V, depending on the voltage drop over the emitter resistors of Q_1, Q_2 and Q_6.

As the name indicates, and as can be seen from (4.3), the PTAT current source is not independent of temperature variations, but ideally displays a variation of

-27%/+33% over the military temperature range from -55 °C to 125 °C. In practice, however, the resistor R_4 also has a temperature coefficient, which for most resistor types is in the order of +500 to +3000 ppm [5]. If an ion-implanted resistor is used, with a temperature coefficient of 3000 ppm, the output current of the PTAT generator can be kept within 5% of its room-temperature value. The absolute value of the PTAT current has the same accuracy as the resistor R_4. This is normally in the order of 10-20%. For a solid design this does not jeopardize the functionality, but if a higher accuracy is needed, for instance in the supply current, an external precision resistor can be used.

4.6 CONCLUSIONS

In the first Section of this Chapter it was shown that the intermediate stage can strongly boost the DC gain of the OpAmp. The design of the intermediate stage, and of the class-AB current control circuit, which was discussed in the second Section, should be compensated for variations in the supply voltage. The feedback-biased class-AB output-current control circuit, introduced in Section 4.2.2, appeared to accurately define the quiescent current through the output transistors, without causing problems in the high-frequency behavior of the output stage. To protect the circuit from heavy saturation of the output transistors, a simple solution was offered in Section 4.3, and in Section 4.4, a new circuit to limit excessive output currents was introduced. Finally, a PTAT reference-current generator was discussed in Section 4.5.

4.7 REFERENCES

[1] J. H. Huijsing, *Integrated Circuits for Accurate Linear Analogue Electrical Signal Processing*. Ph.D. Thesis, Delft University of Technology, Delft, the Netherlands 1981, pp. 81-113.

[2] E. Seevinck, W. de Jager and P. Buitendijk, "A Low-Distortion Output Stage with Improved Stability for Monolithic Power Amplifiers", *IEEE J. Solid-State Circuits*, Vol. SC-23, pp. 794-801, June 1988.

[3] J. Fonderie, M.M. Maris, E. J. Schnitger and J. H. Huijsing, "1-V Operational Amplifier with Rail-to-Rail Input and Output Ranges", *IEEE J. Solid-State Circuits*, Vol. SC-24, pp. 1551-1559, Dec. 1989.

[4] H. C. Nauta and E. H.. Nordholt, "New Class of High-Performance PTAT Current Sources", *Electronic Letters*, Vol. 21, 1985, pp. 384-386.

[5] P. R. Gray and R. G. Meyer, *Analyses and Design of Analog Integrated Circuits*. New York: Wiley, 1984, p. 117.

CHAPTER 5
FREQUENCY COMPENSATION

The input, output and intermediate stages of low-voltage OpAmps, as described in the previous Chapters, all have collector outputs and hence contribute dominating poles to the transfer from input to output. If there are two or more dominating poles, the phase margin of the frequency response is negative, and overall feedback of the OpAmp is not possible. To change this, frequency compensation techniques must be applied that give the OpAmp one dominating pole frequency with a straight 6-dB/octave frequency roll-off. The technique of narrow-banding causes an unacceptable reduction of the bandwidth and is besides difficult to implement in an monolithic integrated circuit. The pole-zero cancelation technique is also unusable because the values of the poles fluctuate with the output current, which would make the cancellation very inaccurate, giving rise to pole-zero doublets and hence slow settling of the step response. The most effective way to compensate the OpAmp is by applying the technique of pole splitting [1]. Simple pole spitting, using one Miller capacitor is discussed in the first Section. The second Section deals with nested Miller compensation for OpAmps consisting of three stages. In the third Section multi-path-driven Miller compensation is introduced. This technique combines the higher bandwidth of the Miller-compensated OpAmp with the higher gain of the nested Miller-compensated OpAmp. Finally, the phenomenon of slewing is the topic of the fourth Section.

5.1 SIMPLE MILLER COMPENSATION

The most simple configuration of an low-voltage OpAmp is a differential input stage as presented in Chapter 2, followed by an output stage, of which several types were discussed in Chapter 3. This OpAmp has initially two dominant pole frequencies, one at the output of the output stage, and one at the output of the input stage. To give the OpAmp a stable frequency response, these poles can be split with a Miller capacitor.

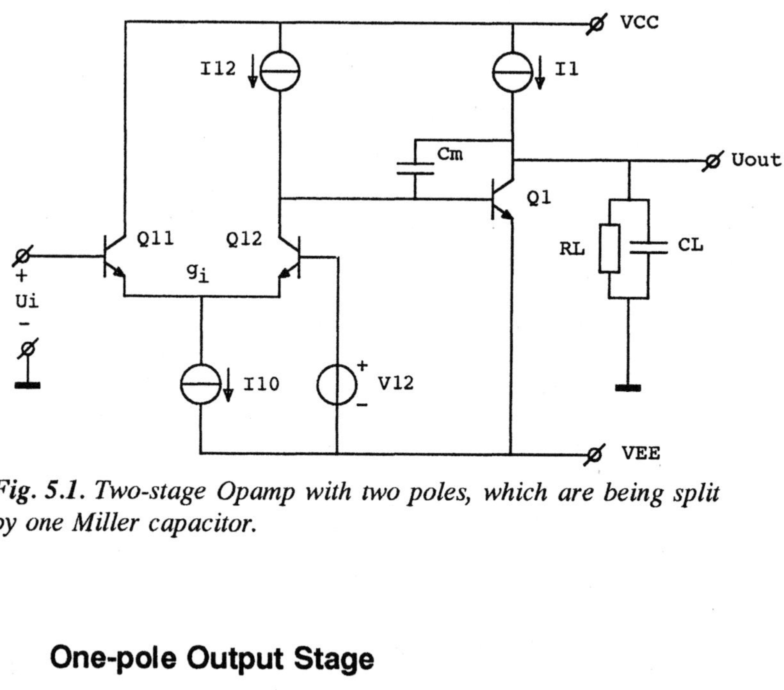

Fig. 5.1. Two-stage Opamp with two poles, which are being split by one Miller capacitor.

5.1.1 One-pole Output Stage

Fig. 5.1 shows the basic configuration of a very simple OpAmp consisting of a differential input stage and an output stage that embodies just one common-emitter (CE) stage. The small-signal equivalent shown in Fig. 5.1 is shown in Fig. 5.2. As discussed in Chapter 3, the equivalent circuit comprises simplified hybrid-π circuits where additional circuit elements, symbolizing parasitic effects, are omitted. In contrast with the schematics presented in Chapter 3, a load resistor

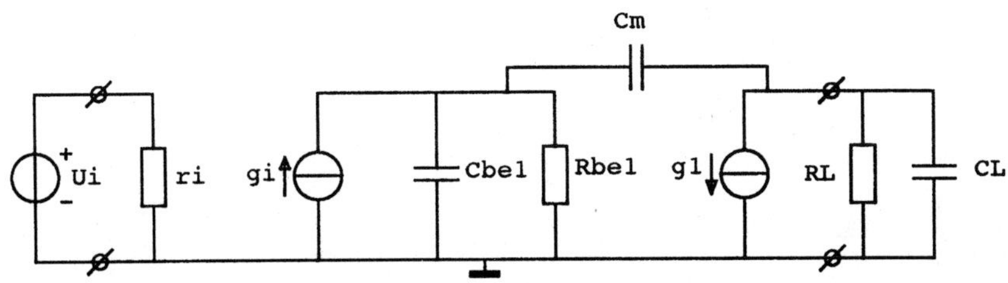

Fig. 5.2. Small-signal equivalent circuit of the Miller compensated two-stage OpAmp shown in Fig. 5.1.

R_L is now assumed parallel to the load capacitor C_L, giving the low-frequency dominant pole a finite value. In this way the movement of the poles as a result of the insertion of C_m can more clearly be illustrated. If the Miller-compensation loop is not yet closed, but C_m is instead connected between the base of Q_1 and ground, the transfer from input voltage to output voltage can be calculated from Fig. 5.2:

$$\frac{U_o}{U_i} = \frac{A_0}{1 + \frac{s}{\omega_1^{(nc)}} + \frac{s^2}{\omega_1^{(nc)}\omega_2^{(nc)}}} \tag{5.1}$$

The superscript *(nc)* indicates that these are the poles of the non-compensated OpAmp. The DC gain A_0 can be calculated:

$$A_0 = -\beta_1 g_i R_L \tag{5.2}$$

The poles are located at the outputs of the two stages. The first pole, at the output of the output stage, is determined by the load resistor and the load capacitor:

$$\omega_1^{(nc)} = \frac{1}{R_L C_L} \tag{5.3}$$

The second pole, at the output of the input stage, is defined by the input resistance of the output stage, and by the sum of the Miller capacitor and the diffusion capacity of Q_1:

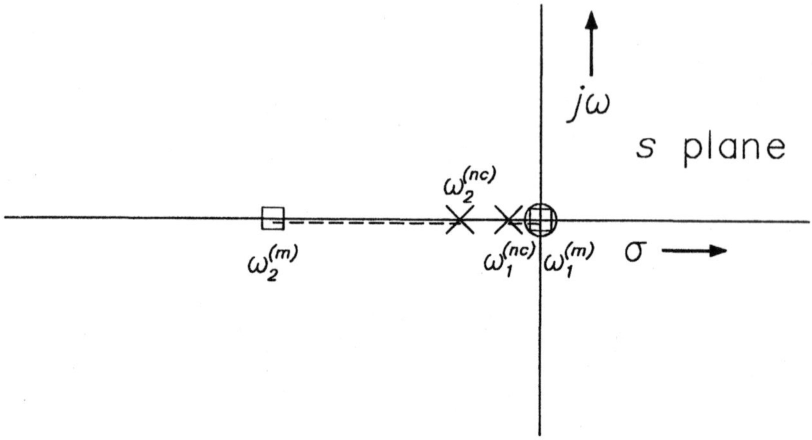

Fig. 5.3. Root locus for the two-stage OpAmp shown in Fig. 5.1. The locus shows the movement of the poles of the loop gain through C_m and the output stage.

$$\omega_2^{(nc)} = \frac{1}{r_{be1}\left(C_{be1} + C_m\right)} \tag{5.4}$$

The gain of the compensation loop through C_m and the output stage has the same poles and an additional zero in the origin, because the loop passes through C_m. If this loop is closed, we can draw the root locus, as is done in Fig. 5.3. The root locus starts at the pole frequencies of the non-compensated OpAmp, indicated with crosses, and it ends at the squares. The zero is symbolized by a circle. The input stage is assumed not to add poles to the transfer of the OpAmp, so the end points of the locus are also the poles of the OpAmp after compensation. These are indicated by the superscript *(m)*, and they can be calculated from Fig. 5.2:

$$\omega_1^{(m)} = \frac{1}{\beta_1 R_L C_m} \tag{5.5}$$

$$\omega_2^{(m)} = \frac{1}{r_{e1} C_L \left(1 + \frac{C_{be1}}{C_m}\right)} = \frac{g_o}{C_L \left(1 + \frac{C_{be1}}{C_m}\right)} \tag{5.6}$$

The Equation of the voltage gain of the OpAmp, after the compensation regains the same shape as before compensation, if the *(nc)* poles are replaced by the *(m)* ones:

$$\frac{U_o}{U_i} = \frac{A_0}{1 + \frac{s}{\omega_1^{(m)}} + \frac{s^2}{\omega_1^{(m)} \omega_2^{(m)}}} \tag{5.7}$$

This is similar to the result derived in Section 3.1, when the voltage gain of the intermediate and output stage was calculated. The only restriction on the value of the transconductance g_i of the intermediate stage then was that the overall transfer should have a positive phase margin. We will now determine g_i more accurately.

If the OpAmp is used as voltage amplifier and over-all feedback is applied, the first and second poles move towards each other. Since the largest movement of the poles shows if the OpAmp is used in a unity-gain feedback configuration, the calculation of g_i should emanate from this assumption; even the unity-gain frequency response should not display peaking. The voltage-follower transfer function of the OpAmp is:

$$\frac{U_o}{U_i} = \frac{1}{1 + \frac{s}{\omega_1^{(u)}} + \frac{s^2}{\omega_1^{(u)} \omega_2^{(u)}}} \tag{5.8}$$

In line with the foregoing, *(u)* now implies that we are dealing with unity-gain feedback applied to the OpAmp. The first coefficient, $\omega_1^{(u)}$, is calculated by solving the Kirchhoff current equations of the circuit of Fig. 5.2 when unity-gain feedback is applied to it:

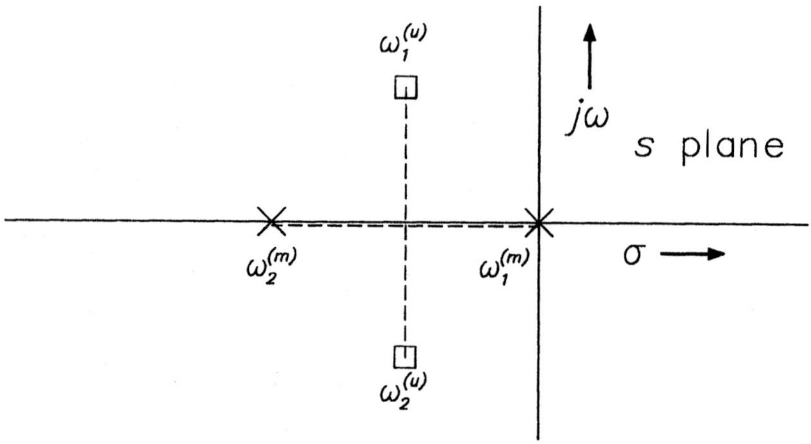

Fig. 5.4. Root locus for the two-stage OpAmp. The locus shows the movement of the poles if unity-gain feedback is applied to the Miller-compensated OpAmp.

$$\omega_1^{(u)} = \frac{g_i}{C_m} \tag{5.9}$$

and the second coefficient, $\omega_2^{(u)}$, is:

$$\omega_2^{(u)} = \omega_2^{(m)} \tag{5.10}$$

Since the two coefficients of (5.8) are of the same order of magnitude, the pole positions in the s-plane are not given by (5.9) and (5.10), but instead a complex pair is formed. The value of g_i should now be chosen in such a way that the poles are not too complex, but so that (5.8) still does have the largest possible -3-dB bandwidth. In other words, (5.8) should have the maximally-flat-magnitude or Butterworth frequency characteristic. The second-order Butterworth transfer has the form:

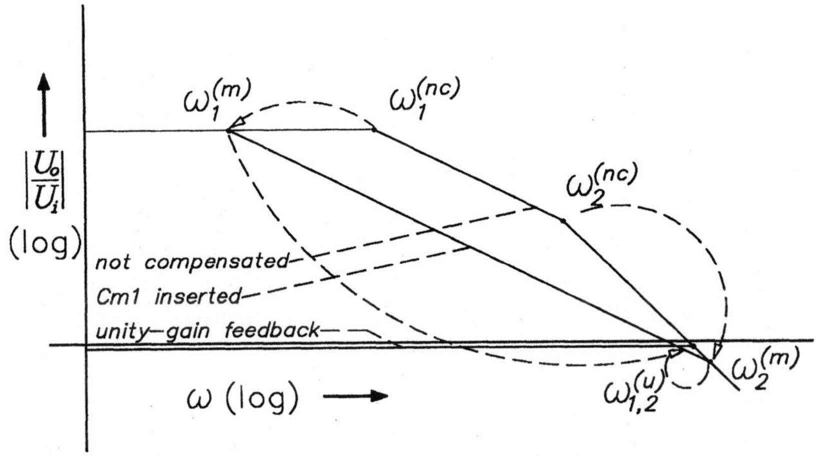

Fig. 5.5. Frequency response of the OpAmp shown in Fig. 5.1. Indicated are: the uncompensated response (top), the Miller-compensated response (middle), and the unity-gain or voltage-follower response (bottom).

$$\frac{U_o}{U_i} = \frac{1}{\left(1 + \dfrac{s}{\omega_0^{(u)} e^{+j45°}}\right)\left(1 + \dfrac{s}{\omega_0^{(u)} e^{-j45°}}\right)} \tag{5.11}$$

where $\omega_0^{(u)}$ is the -3-dB bandwidth of the voltage follower, in radians per second, and $e^{+j\phi}$ is the unity vector in the s-plane with an angle ϕ with the negative σ-axis. Equating (5.8) and (5.11) gives:

$$\omega_0^{(u)} = \sqrt{2}\,\omega_1^{(u)} = \sqrt{2}\,\frac{g_i}{C_m} \tag{5.12}$$

and:

$$\omega_2^{(u)} = 2\omega_1^{(u)} \tag{5.13}$$

The root locus depicted in Fig. 5.4 shows the movement of the poles from the Miller-compensated positions to the Butterworth positions. The root locus starts at the poles of (5.7), given in (5.5) and (5.6), and it ends at the poles of (5.8). Returning to the open-loop configuration, the open-loop bandwidth $\omega_0^{(m)}$ of the OpAmp can be found with (5.5), (5.7) and 5.12):

$$\omega_0^{(m)} = A_0 \omega_1^{(m)} = \frac{g_i}{C_m} = \frac{1}{2}\omega_2^{(m)} = \frac{1}{2}\frac{g_o}{C_L\left(1+\frac{C_{be1}}{C_m}\right)} \tag{5.14}$$

or in words: the ratio of the transconductance of the input stage and the Miller capacitor should be chosen such that the open-loop bandwidth is half the value of the second pole in the open-loop transfer. The voltage-follower function then has the maximal bandwidth, without peaking.

Table 5.1. *Calculated pole positions for a two-stage OpAmp with an n-p-n output stage.*

	non-compensated $\omega^{(nc)}$	*Miller compensated* $\omega^{(m)}$	*unity-gain feedback* $\omega^{(u)}$
ω_0	*20 MHz*	*15 MHz*	*21 MHz*
ω_1	*160 kHz*	*16 kHz*	*21 MHz, +45°*
ω_2	*2.9 MHz*	*29 MHz*	*21 MHz, -45°*
A_0	*59 dB*	*59 dB*	*0 dB*

Fig. 5.5 shows the effect of the pole movements on the frequency response of the OpAmp. First the poles are split from the *(nc)* positions to the *(m)* positions and if then unity-gain feedback is applied to the OpAmp, the poles again move towards each other and coincide at the frequency $\omega_0^{(u)}$. To illustrate the theory we will apply it to an OpAmp with an output stage consisting of 5-GHz n-p-n transistors with a current gain of 100. If the quiescent current through the output

transistor is 500 μA, g_o is 19 mmho and C_{be1} is 0.6 pF. The load consists of a 100 pF capacitor in parallel with a 10 kΩ resistor, and the Miller capacitor is chosen 10 pF. Table 5.1 can now be filled out by solving (5.1), (5.7) and (5.8) using the Equations introduced in this Section. Also from these Equations it is found that the transconductance g_i must be 0.9 mmho, which means that the tail current I_{10} in Fig. 5.1 is 95 μA. It is interesting to see from Fig. 5.5 and Table 5.1 that if unity-gain feedback is applied to the OpAmp, all the 59 dB gain is "transformed" into the maximal obtainable bandwidth, i.e. 21 MHz, the same as the bandwidth of the non-compensated OpAmp.

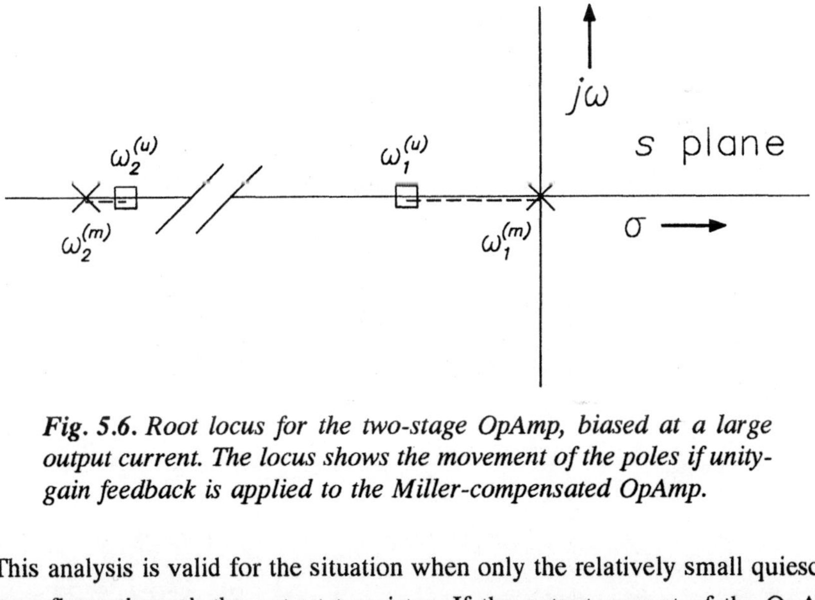

Fig. 5.6. Root locus for the two-stage OpAmp, biased at a large output current. The locus shows the movement of the poles if unity-gain feedback is applied to the Miller-compensated OpAmp.

This analysis is valid for the situation when only the relatively small quiescent current flows through the output transistor. If the output current of the OpAmp increases, the second pole of the open-loop transfer (5.7) moves to a higher frequency. This does not affect the open-loop bandwidth $\omega_0^{(m)}$, but the unity-gain bandwidth $\omega_0^{(u)}$ is smaller than indicated by (5.12), because the poles of the unity-gain transfer are no longer in the Butterworth position. In fact, $\omega_0^{(u)}$ is now equal to $\omega_0^{(m)}$. Because the poles of the unity-gain transfer are now widely separated, (5.11), (5.12) and (5.13) do not apply and the roots of (5.8) are directly given by (5.9) and (5.10). Fig. 5.6 shows the root locus of the OpAmp, biased at a large output current, if unity-gain feedback is applied.

5.1.2 Two-pole Output Stage

Most output stages have more than one pole in their transfer from input current to output voltage. It is likely that this influences the compensation as discussed in the previous section. As an example, an OpAmp consisting of an differential input stage and a Darlington output stage, as depicted in Fig. 5.7, will be analyzed. With the small-signal equivalent circuit of the Darlington output stage, that already has been given in Fig. 3.7, an equivalent circuit similar to that shown in Fig. 5.2 can be used for the calculations.

The voltage gain of the non-compensated OpAmp can be calculated from such an equivalent circuit:

$$\frac{U_o}{U_i} = \frac{A_0}{1 + \frac{s}{\omega_1^{(nc)}} + \frac{s^2}{\omega_1^{(nc)}\omega_2^{(nc)}} + \frac{s^3}{\omega_1^{(nc)}\omega_2^{(nc)}\omega_3^{(nc)}}} \tag{5.15}$$

The DC gain A_0 is:

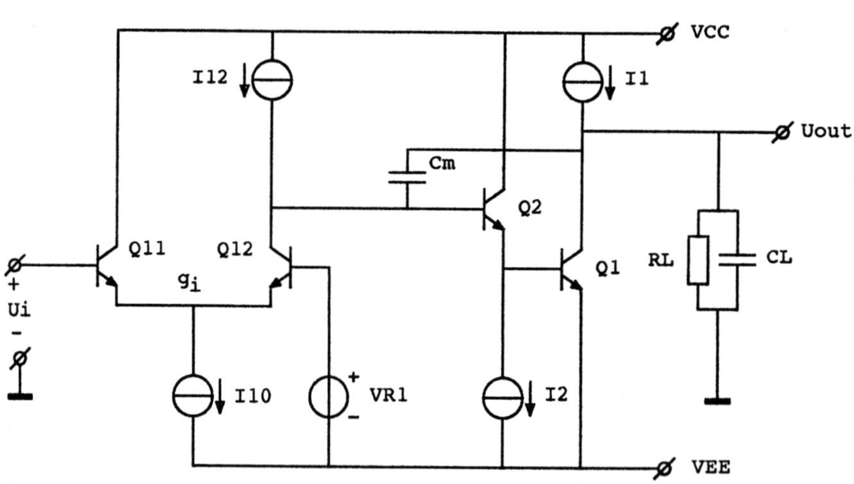

Fig. 5.7. *Two-stage Opamp with one Miller capacitor, and three poles in the transfer from input voltage to output voltage.*

$$A_0 = -\beta_1 \beta_2 g_i R_L \tag{5.16}$$

and the first pole, located at the output is again:

$$\omega_1^{(nc)} = \frac{1}{R_L C_L} \tag{5.17}$$

The second pole, at the output of the input stage, is determined by the input resistance of the output stage and C_m:

$$\omega_2^{(nc)} = \frac{1}{\beta_2 \left(r_{e2} + r_{be1} \right) C_m} \tag{5.18}$$

and the third pole is the internal pole of the Darlington:

$$\omega_3^{(nc)} = \frac{1}{r_p C_{be1}} \tag{5.19}$$

Resistance r_p is equal to r_{e2} and r_{be1} in parallel, as given in (3.6). Now the loop through C_m is closed, and the root locus starts at the *(nc)* poles and ends at the poles of the compensated OpAmp, indicated by *(m)*. The gain is then equal to (5.15), if the coefficients are replaced by:

$$\omega_1^{(m)} = \frac{1}{\beta_1 \beta_2 R_L C_m} \tag{5.20}$$

$$\omega_2^{(m)} = \frac{g_o}{C_L \left(1 + \dfrac{C_{be2}}{C_m} \right)} \tag{5.21}$$

The transconductance of the Darlington output stage, g_o is given in (3.5). The third coefficient is not affected by the Miller compensation:

$$\omega_3^{(m)} = \omega_3^{(nc)} = \frac{1}{r_p C_{bel}} \tag{5.22}$$

To find the appropriate value for g_i, the OpAmp is connected as a voltage follower again, with a unity-gain transfer:

$$\frac{U_o}{U_i} = \frac{1}{1 + \frac{s}{\omega_1^{(u)}} + \frac{s^2}{\omega_1^{(u)}\omega_2^{(u)}} + \frac{s^3}{\omega_1^{(u)}\omega_2^{(u)}\omega_3^{(u)}}} \tag{5.23}$$

If the third pole, the internal pole of the Darlington output stage, is situated at a frequency much higher than the second pole, the transfer can be approximated by a two-pole representation. To obtain the maximal -3dB bandwidth, the poles

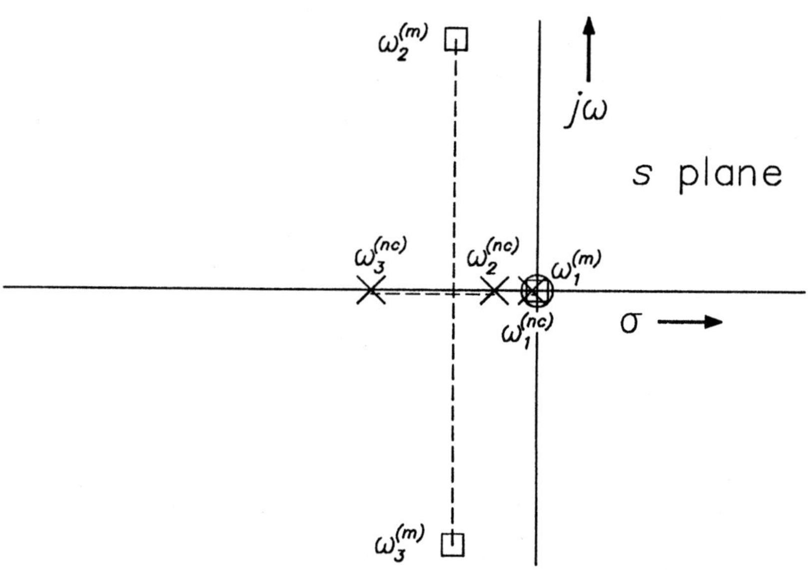

Fig. 5.8. Root locus of the two-stage OpAmp with complex poles in the output stage. The locus shows the movement of the poles of the loop gain through C_m and the output stage. C_{m0} is not inserted.

should then be moved to the second-order Butterworth positions, as shown in the previous section. The poles are then given by (5.12) and (5.13), and the root loci of the Miller compensation, and the unity-gain feedback, are identical to Figs. 5.3 and 5.4, respectively, with the addition of the high-frequency third pole. The open-loop bandwidth is still half the value of the second pole after compensation:

$$\omega_0^{(m)} = \frac{g_i}{C_m} = \frac{1}{2}\omega_2^{(m)} = \frac{1}{2}\frac{g_o}{C_L} \tag{5.24}$$

At higher output-current levels, however, a different root locus emerges. After all, the second pole of the open-loop transfer then moves to higher frequencies, and

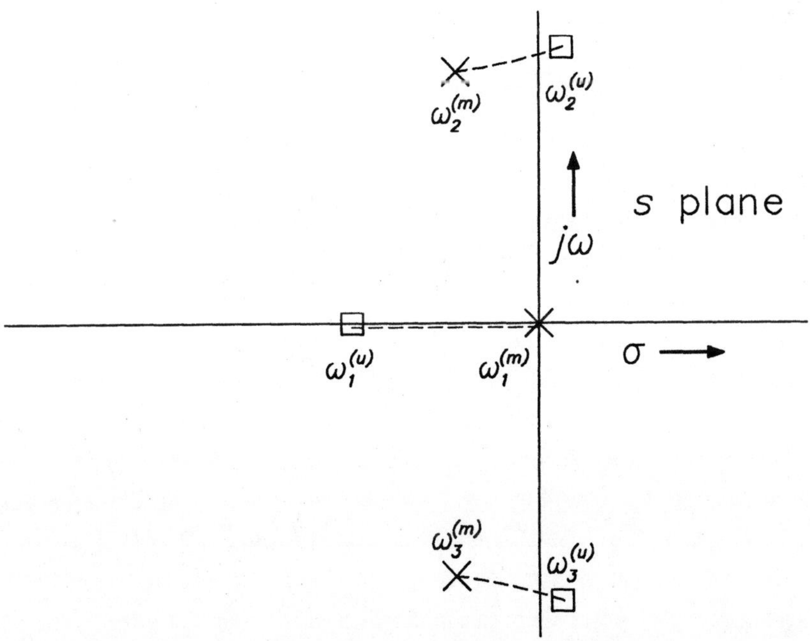

Fig. 5.9. The root locus of the two-stage OpAmp with complex poles in the output stage. The locus shows the movement of the poles if unity-gain feedback is applied to the compensated OpAmp. C_{m0} is not inserted.

Table 5.2. Calculated pole positions for an two-stage OpAmp with a p-n-p Darlington output stage biased at 500 μA, loaded with 100 pF in parallel with 10 kΩ and with a C_{m0} of 20 pF.

	non-compensated $\omega^{(nc)}$	Miller compensated $\omega^{(m)}$	unity-gain feedback $\omega^{(u)}$
ω_0	4.9 MHz	1.6 MHz	2.2 MHz
ω_1	32 kHz	4 kHz	2.2 MHz, +45°
ω_2	380 kHz	3.1 MHz	2.2 MHz, -45°
ω_3	25 MHz	25 MHz	25 MHz
A_0	52 dB	52 dB	0 dB

eventually forms a complex pair with the third pole. Because the complex poles are situated at high frequencies, the overall gain is already much smaller than one and the influence of the complex poles is relatively small.

If the frequency where the third pole in the open-loop transfer is located is not much higher than $\omega_0^{(m)}$ and $\omega_0^{(u)}$, the influence of complex poles that arise when the output current increases is larger. This is shown in Figs. 5.8 and 5.9. Suppose the value of g_i is determined by (5.24), that is, as if the frequency behavior of the output stage were dominated by one, and not two poles. When unity-gain feedback is applied, the two poles become even more complex and eventually move to the left-hand plane, as shown in Fig. 5.9. To avoid this instability, an additional Miller capacitor C_{m0} should be inserted, as was argued in Section 3.2. The capacitor C_{m0} is chosen large enough to make the damping ratio at large output currents at least 0.7, giving complex poles with an angle of no more than 45° with the negative σ-axis. At the quiescent current, ω_3, given in (3.15), is now situated at a much higher frequency and the transfer can again be approximated by a two-pole representation. The unity-gain bandwidth $\omega_0^{(m)}$ is fixed at half the new, and lower, value of $\omega_2^{(m)}$, given in (3.14). An output stage consisting of 50-MHz p-n-p's with a β of 20 is taken as an example. The load is a 10 kΩ resistor in parallel with a capacitor of 100 pF and the Miller capacitor is 10 pF. Without C_{m0}, the OpAmp would be unstable if overall feedback were applied, and therefore also a C_{m0} of 20 pF is inserted. At a quiescent current of 500 μA, and with I_2 set to zero, r_{e1} is 52 Ω, C_{be1} is 61 pF and r_{e2} is 1.0 kΩ. For simplicity, C_{be2} is neglected in the calculations. By solving (5.23) with the coefficients given in (5.16) through (5.21), Table 5.2 can now be filled. To get a transconductance g_i=98 μmho that satisfies (5.24), the current source I_{10} is 10 μA. With C_{m0}, $\omega_2^{(m)}$ is lowered, and therefore the bandwidth is also reduced, but the instability and peaking at high current levels are effectively avoided. If the p-n-p Darlington is compared with the single n-p-n output transistor that appeared on the scene earlier, we find that although both output stages are biased at the same quiescent current, the bandwidth of the p-n-p Darlington is nine times lower. This once more demonstrates the importance of output stages with only one dominating pole frequency as was emphasized in Chapter 3. The compensation of OpAmps comprising a one-pole output stage is easier to design and it also yields a higher bandwidth.

5.2 NESTED MILLER COMPENSATION

The OpAmps shown in Figs. 5.1 and 5.7 have a DC gain that is rather low if they are to serve as general-purpose devices. The presented examples showed this. That is why an intermediate stage can be placed between the input and output stage. In the case of low-voltage OpAmps, this intermediate stage is a differential pair which therefore has a collector output. This collector output adds an extra pole to the transfer function of the OpAmp, and a more complex compensation technique must be used.

Consider the OpAmp as described in the previous Section, which consists of an output stage and a differential pair, in this case the intermediate stage. After the Miller compensation, we could look upon these two stages as if they were one compound output stage, with one (low-frequency) dominating pole. If then an input stage is placed in front of this "output stage", and a new pole is added, the frequency compensation can be repeated. This technique of repeated frequency compensation is called nested Miller compensation [2], [3], [4]. In line with the preceding, the nested Miller compensation is first applied to an OpAmp comprising a one-pole output stage.

5.2.1 One-pole Output Stage

Fig. 5.10 shows a three-stage OpAmp, comprising an input stage, an intermediate stage and an n-p-n output transistor. The intermediate stage and the output transistor are compensated with Miller capacitor C_{m1}, and the combination of these two, together with the input stage are in turn compensated with C_{m2}. The small-circuit equivalent is shown in Fig. 5.11.

The overall transfer from input voltage to output voltage, with C_{m1} and C_{m2} still connected to ground, can be calculated from Fig. 5.11:

$$\frac{U_o}{U_i} = \frac{A_0}{1 + \frac{s}{\omega_1^{(nc)}} + \frac{s^2}{\omega_1^{(nc)}\omega_2^{(nc)}} + \frac{s^3}{\omega_1^{(nc)}\omega_2^{(nc)}\omega_3^{(nc)}}} \tag{5.25}$$

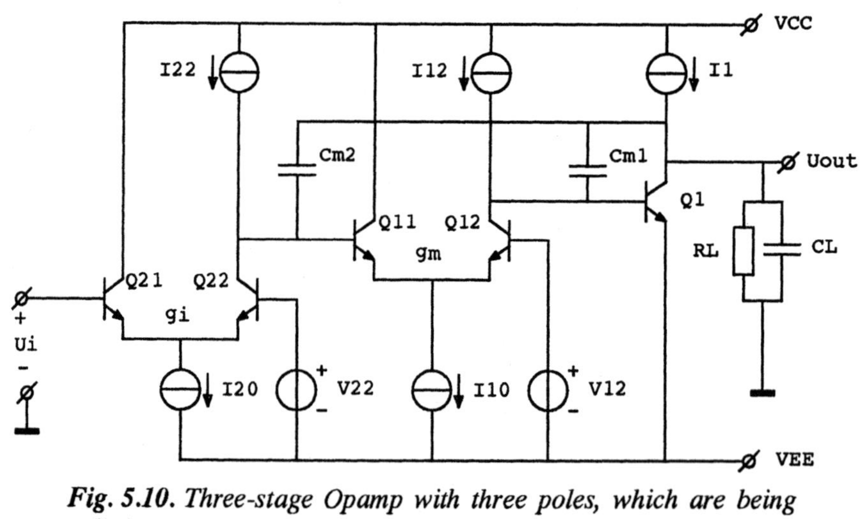

Fig. 5.10. Three-stage Opamp with three poles, which are being split by two nested Miller capacitors.

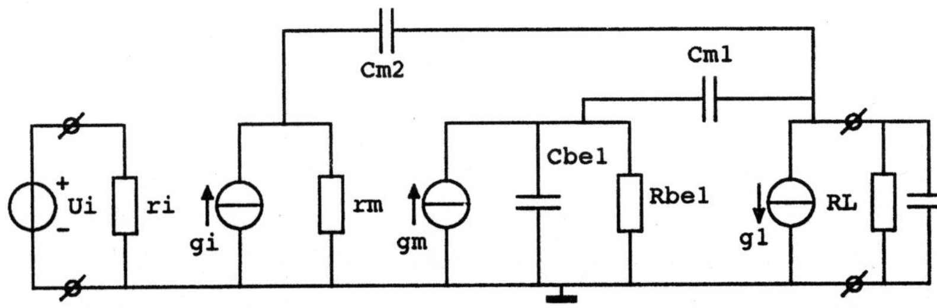

Fig. 5.11. Small-signal equivalent circuit of the three-stage OpAmp of Fig. 5.10, with two nested Miller capacitors.

If β_m is the current gain of the intermediate-stage transistors and g_m is the transconductance of the input stage, the DC voltage gain is:

$$A_0 = -\beta_m \beta_1 g_i R_L \tag{5.26}$$

The first pole is defined by the load resistor and capacitor:

$$\omega_1^{(nc)} = \frac{1}{R_L C_L} \tag{5.27}$$

The second pole is situated at the output of the input stage and is determined by the input resistance of the intermediate stage and C_{m2}:

$$\omega_2^{(nc)} = \frac{g_m}{\beta_m C_{m2}} = \frac{1}{r_{im} C_{m2}} \tag{5.28}$$

where r_{im} is the input resistance and g_m the transconductance of the intermediate stage. The third pole, at the output of the intermediate stage, is controlled by the input resistance of the output stage and the capacitance at this input:

$$\omega_3^{(nc)} = \frac{1}{r_{be1} C_{m1}\left(1 + \frac{C_{be1}}{C_{m1}}\right)} \tag{5.29}$$

As before, the superscript *(nc)* indicates that these are the poles of the non-compensated OpAmp. The superscripts *(m1)*, *(m2)* and *(u)* in the Equations that follow make it possible to distinguish the poles of the first Miller loop, the second Miller loop and the unity-gain feedback loop, respectively.

To see how the poles of this transfer move after the insertion of the Miller capacitors, we first apply the analysis of the two-stage OpAmp presented in the previous Section to the intermediate stage and output transistor. The poles of the loop gain through C_{m1} and Q_1, are equal to $\omega_1^{(nc)}$ and $\omega_3^{(nc)}$, given above.
If the loop around the output stage is closed, the poles move to the positions given by:

$$\omega_1^{(m1)} = \frac{1}{\beta_1 R_L C_{m1}} \tag{5.30}$$

and:

$$\omega_3^{(m1)} = \frac{1}{r_{e1} C_L \left(1 + \frac{C_{be1}}{C_{m1}}\right)} = \frac{g_o}{C_L \left(1 + \frac{C_{be1}}{C_{m1}}\right)} \tag{5.31}$$

The root locus of this first Miller splitting is illustrated in the top-most part of Fig. 5.12. The frequency response of the intermediate and output stage, before and after insertion of C_{m1}, is shown in the upper part of Fig. 5.13.

The overall transfer of the OpAmp is of course also affected by this first compensation. The expression of the voltage gain is similar to (5.25), when the *(nc)* superscripts are replaced by *(m1)*. The first and third pole are now given by (5.30) and (5.31), and the pole at the output of the input stage, $\omega_2^{(m1)}$, is still in the position given by (5.28).

Now we can compute the loop gain through C_{m2}, the intermediate and output stage. The poles of this loop gain are once more (5.30), (5.28) and (5.31). When the loop is closed, the second Miller capacitor splits the first and second pole and keeps the third one unaltered:

$$\omega_1^{(m2)} = \frac{1}{\beta_m \beta_1 R_L C_{m2}} \tag{5.32}$$

$$\omega_2^{(m2)} = \frac{g_m}{C_{m1}} \tag{5.33}$$

$$\omega_3^{(m2)} = \omega_3^{(m1)} \tag{5.34}$$

The root locus of the loop gain through C_{m2} is shown in the middle part of Fig. 5.12. The overall transfer of the OpAmp after C_{m2} is inserted is again of the form of (5.25), now with the poles (5.32), (5.33) and (5.34). The overall frequency response, before compensation, after insertion of C_{m1} and after insertion of C_{m2} is shown in the lower part of Fig. 5.13.

To determine the value of the input and intermediate stage transconductance, unity-gain feedback is applied to the OpAmp, just as it was done with respect to the two-stage OpAmp. The unity-gain transfer again has the form of (5.25), but now A_0 is of course +1, and the coefficients are:

$$\omega_1^{(u)} = \frac{g_i}{C_{m2}} \tag{5.35}$$

$$\omega_2^{(u)} = \frac{g_m}{C_{m1}} \tag{5.36}$$

and:

$$\omega_3^{(u)} = \omega_3^{(m2)} = \omega_3^{(m1)} = \frac{g_o}{C_L\left(1 + \frac{C_{be1}}{C_{m1}}\right)} \tag{5.37}$$

Because the poles are close together (5.35), (5.36) and (5.37) are not the roots of (5.25), but merely its coefficients. To obtain the maximal bandwidth, without peaking, the poles should be in third-order Butterworth positions, given by:

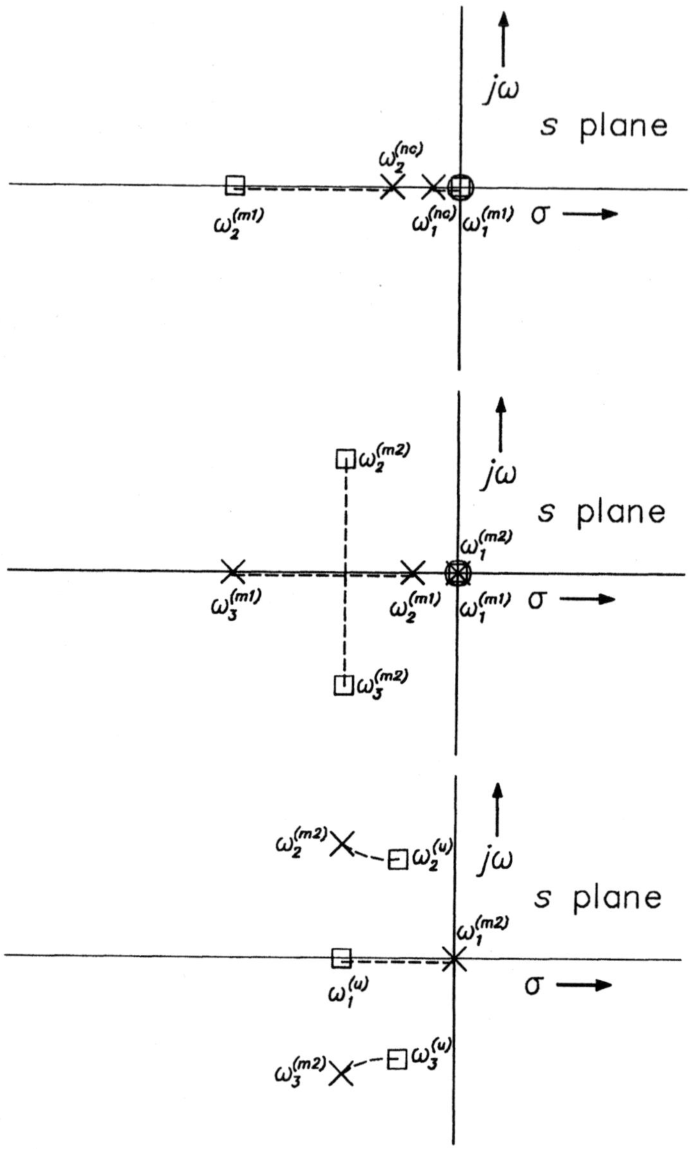

Fig. 5.12. Root loci of: loop gain through C_{m1} and the output stage (top), loop gain through C_{m2}, the intermediate and the output stage (middle), the OpAmp, if unity-gain feedback is applied (bottom).

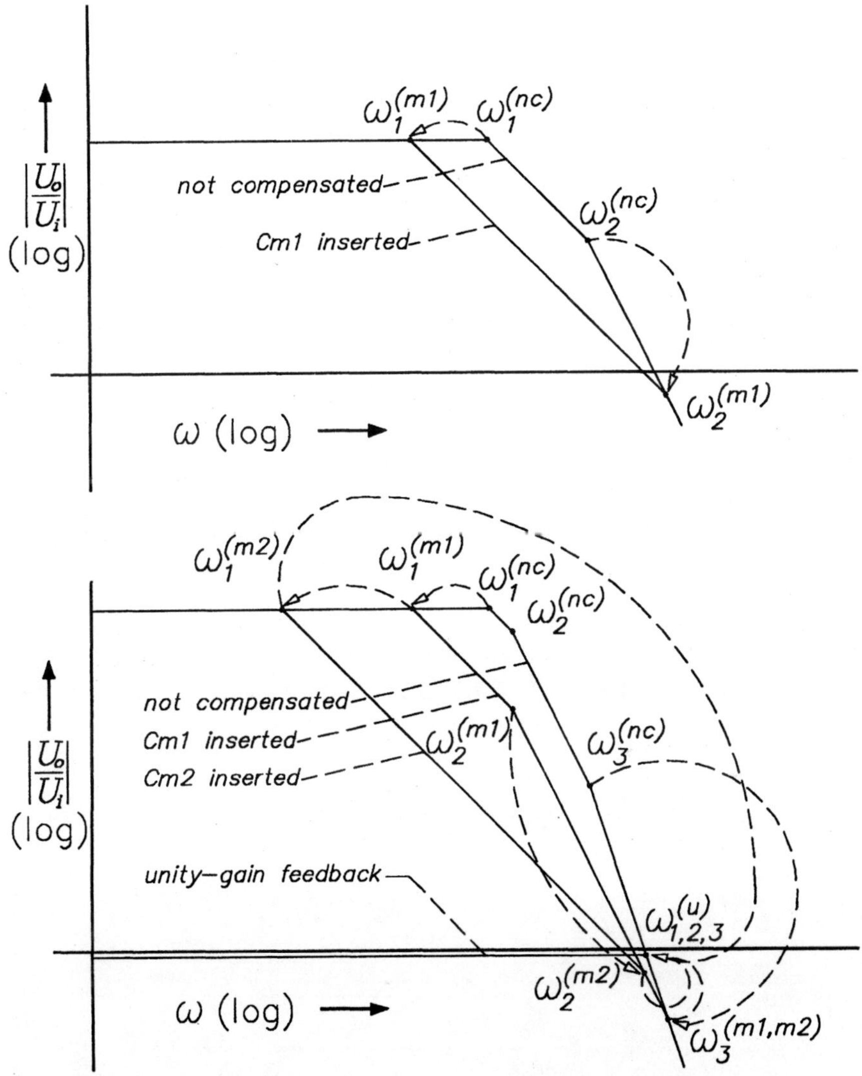

Fig. 5.13. *Frequency response of the three-stage OpAmp. On top the movement of the poles of the output and intermediate stage, below the movement of the poles of the complete OpAmp.*

$$\frac{U_o}{U_i} = \frac{1}{\left(1 + \frac{s}{\omega_0^{(u)} e^{+j60°}}\right)\left(1 + \frac{s}{\omega_0^{(u)}}\right)\left(1 + \frac{s}{\omega_0^{(u)} e^{+j60°}}\right)} \tag{5.38}$$

Equalizing the unity-gain transfer and (5.38) gives:

$$\omega_0^{(u)} = 2\,\omega_1^{(u)} = 2\frac{g_i}{C_{m2}} \tag{5.39}$$

$$\omega_2^{(u)} = 2\omega_1^{(u)} \tag{5.40}$$

$$\omega_3^{(u)} = 2\omega_2^{(u)} = 4\,\omega_1^{(u)} \tag{5.41}$$

The lower part of Fig. 5.12 shows the root locus of the unity-gain feedback. The poles start at the *(m2)* positions, the poles of the open-loop transfer, and they end at the *(u)* positions. As could be expected, these last poles are equally spaced in the left-hand plane. The unity-gain frequency response is depicted in the lowest graph of the lower part of Fig. 5.13. The open-loop bandwidth, in relation to the transconductance of the input and intermediate stage, is now also fixed:

$$\omega_0^{(m2)} = \frac{g_i}{C_{m2}} = \frac{1}{4}\frac{g_o}{C_L\left(1 + \frac{C_{be1}}{C_{m1}}\right)} \tag{5.42}$$

and:

$$\omega_2^{(m2)} = \frac{g_m}{C_{m1}} = \frac{1}{2}\frac{g_o}{C_L\left(1 + \frac{C_{be1}}{C_{m1}}\right)} \tag{5.43}$$

The transconductance of the intermediate stage should be chosen such that the value of $\omega_2^{(m2)}$ is half the value of $\omega_3^{(m2)}$. This is identical to the result derived in the previous Section. The transconductance of the input stage should be chosen

Table 5.3. Coefficients of the characteristic equation of the three-stage OpAmp with one output transistor.
"(nc)"= non-compensated,
"(m1)"= after insertion C_{m1},
"(m2)"= after insertion C_{m2},
"(u)"= after unity-gain feedback.

	$\omega^{(nc)}$	$\omega^{(m1)}$	$\omega^{(m2)}$	$\omega^{(u)}$
ω_0	$\sqrt[3]{\dfrac{g_i g_m g_o}{C_{m2} C_{m1} C_L \left(1 + \dfrac{C_{be1}}{C_{m1}}\right)}}$	$\sqrt{\dfrac{g_i g_m}{C_{m2} C_{m1}}}$	$\dfrac{g_i}{C_{m2}}$	$\dfrac{2 g_i}{C_{m2}}$
ω_1	$\dfrac{1}{R_L C_L}$	$\dfrac{1}{\beta_1 R_L C_{m1}}$	$\dfrac{1}{\beta_m \beta_1 R_L C_{m2}}$	$\dfrac{g_i}{C_{m2}}$
ω_2	$\dfrac{1}{r_{im} C_{m2}}$	$\dfrac{1}{r_{im} C_{m2}}$	$\dfrac{g_m}{C_{m1}}$	$\dfrac{g_m}{C_{m1}}$
ω_3	$\dfrac{1}{r_{be1}\left(C_{be1} + C_{m1}\right)}$	$\dfrac{g_o}{C_L\left(1 + \dfrac{C_{be1}}{C_{m1}}\right)}$	$\dfrac{g_o}{C_L\left(1 + \dfrac{C_{be1}}{C_{m1}}\right)}$	$\dfrac{g_o}{C_L\left(1 + \dfrac{C_{be1}}{C_{m1}}\right)}$

such that the open-loop bandwidth $\omega_0^{(m2)}$ of the OpAmp is one quarter of the value of the third pole in the open-loop transfer $\omega_3^{(m2)}$. The two poles $\omega_2^{(m2)}$ and $\omega_3^{(m2)}$ are then in second-order Butterworth position. If unity-gain feedback is applied to the OpAmp, all three poles move into the third-order Butterworth position.

To summarize all information about the poles of the three-stage OpAmp, Table 5.3 gives the expressions for the coefficients of (5.25). Table 5.4 shows the same information, but now with values found in the following example. The output stage described in the example at the end of Section 5.1.1 is again used:

Table 5.4. Calculated pole positions for a three-stage OpAmp with an n-p-n output stage, biased at 500 μA and loaded with 100 pF in parallel with 10 kΩ.

	non-compensated $\omega^{(nc)}$	Miller compensated $\omega^{(m1)}$	twice Miller compensated $\omega^{(m2)}$	unity-gain feedback $\omega^{(u)}$
ω_0	15 MHz	10 MHz	7.3 MHz	15 MHz
ω_1	160 kHz	16 kHz	320 Hz	15 MHz, +60°
ω_2	290 kHz	290 kHz	21 MHz, +45°	15 MHz, 0°
ω_3	2.9 MHz	29 MHz	21 MHz, -45°	15 MHz, -60°
A_0	87 dB	87 dB	87 dB	0 dB

the transistors are 5-GHz n-p-n's with a β of 100, R_L is 10 kΩ, C_L is 100 pF, C_{m1} is 10 pF and, furthermore, C_{m2} is 5 pF. The characteristic equation, the denominator of (5.25), is then solved with the coefficients given in Table 5.3. The transconductance of the intermediate stage, g_m, is now 0.9 mmho and g_i is 230 μmho. The voltage follower again "transforms" the 87 dB gain of the OpAmp into a -3-dB bandwidth of 15 MHz, equal to the bandwidth of the non-compensated OpAmp. If the bandwidth of this three-stage OpAmp is compared with that of the two-stage OpAmp, we see that the increase of the DC gain has to be paid for with a reduction of the bandwidth by a factor of two.

The starting point of the assumption that the unity-gain bandwidth $\omega_0^{(m2)}$ is determined only by the transconductance of the input stage and C_{m2}, is that the higher-order poles $\omega_2^{(m2)}$ and $\omega_3^{(m2)}$ are situated at a frequency high enough not to influence the bandwidth. Table 5.4 shows that the absolute value of the Butterworth poles $\omega_2^{(m2)}$ and $\omega_3^{(m2)}$ is about three times that of the bandwidth. Including the Butterworth poles at a three times higher frequency in the calculation of $\omega_0^{(m2)}$, gives a deviation of only 4%. The assumption therefore seems valid. If the output current of the OpAmp increases, however, the third-order pole moves

Table 5.5. Calculated pole positions for a three-stage OpAmp with an n-p-n output stage, biased at 5 mA and loaded with 100 pF in parallel with 10 kΩ.

	non-compensated $\omega^{(nc)}$	Miller compensated $\omega^{(m1)}$	twice Miller compensated $\omega^{(m2)}$	unity-gain feedback $\omega^{(u)}$
ω_0	27 MHz	10 MHz	6.0 MHz	10 MHz
ω_1	160 kHz	16 kHz	320 Hz	10 MHz, +45°
ω_2	290 kHz	290 kHz	15 MHz	10 MHz, -45°
ω_3	19 MHz	190 MHz	190 MHz	190 MHz
A_0	87 dB	87 dB	87 dB	0 dB

to a higher frequency, similar to what is discussed at the end of Section 5.1.1. The second pole $\omega_2^{(m2)}$ is now only about two times the bandwidth. If we calculate the bandwidth again without neglecting the influence of this single pole, it is found that the lowering is now 17%, which is considerable. This explains the in-practice observation of the lowering of the open-loop unity-gain bandwidth at higher output currents. For purpose of comparison, Table 5.5 shows the poles of the OpAmp from the example above, but now at a current through the output transistor of 5 mA. The open-loop gain no longer has a complex pair of poles and the unity-gain transfer has second-order instead of third-order Butterworth poles.

5.2.2 Two-pole Output Stage

The Equations given in the preceding Section consider an OpAmp with an output stage that has only one dominant pole frequency. In Section 5.1.2 it was shown that this hypothesis is not always legitimate, especially not if the output stage consists of p-n-p transistors. Fig. 5.14 shows a three-stage OpAmp with a

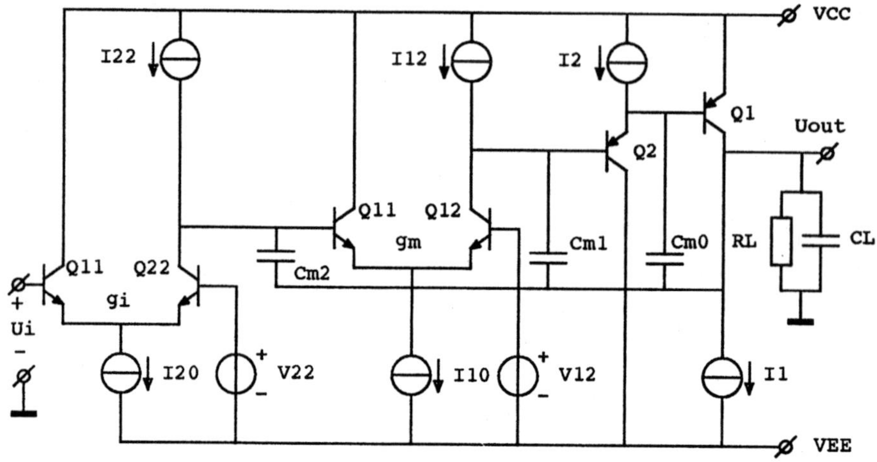

Fig. 5.14. Three-stage OpAmp with four poles, comprising a Darlington output stage, and two nested Miller capacitors.

p-n-p Darlington output stage. To avoid instability deriving from the complex poles of the output stage, C_{m0} is also inserted. The small-signal equivalent circuit can be put together from the circuits shown earlier in this Chapter. First, the non-compensated frequency response, with the Miller capacitors C_{m1} and C_{m2} connected to ground, is again analyzed. This transfer now has a fourth-order pole as well:

$$\frac{U_o}{U_i} = \frac{A_0}{1 + \frac{s}{\omega_1^{(nc)}} + \frac{s^2}{\omega_1^{(nc)}\omega_2^{(nc)}} + \frac{s^3}{\omega_1^{(nc)}\omega_2^{(nc)}\omega_3^{(nc)}} + \frac{s^4}{\omega_1^{(nc)}\omega_2^{(nc)}\omega_3^{(nc)}\omega_4^{(nc)}}} \tag{5.44}$$

The gain A_0 is larger than it was for the output stage consisting of just one transistor:

$$A_0 = -\beta_m \beta_1 \beta_2 g_i R_L \tag{5.45}$$

The first pole is still defined by the load resistor and capacitor:

$$\omega_1^{(nc)} = \frac{1}{R_L\left(C_L + \beta_1 C_{m0}\right)} \tag{5.46}$$

The second pole is situated at the output of the input stage and is determined by the input resistance of the intermediate stage and C_{m2}:

$$\omega_2^{(nc)} = \frac{g_m}{\beta_m C_{m2}} = \frac{1}{r_{im} C_{m2}} \tag{5.47}$$

where once more r_{im} is the input resistance and g_m the transconductance of the intermediate stage. If the base-emitter capacitance of Q_2 is neglected, the third-order-pole is determined by the input resistance of the output stage and C_{m1}:

$$\omega_3^{(nc)} = \frac{1}{\beta_2\left(r_{e2} + r_{be1}\right) C_{m1}} \tag{5.48}$$

Finally, the internal pole of the output stage is located at the base of Q_1 and is equal to:

$$\omega_4^{(nc)} = \frac{1 + \dfrac{\beta_1 C_{m0}}{C_L}}{r_p C_{be1}} \tag{5.49}$$

If the Miller capacitors C_{m1} and C_{m2} are connected from the output to the inputs of the output and intermediate stage, respectively, the poles are split again, and the transfer from input voltage to output voltage can be calculated, as in the previous Sections. The *(nc)*, *(m1)*, *(m2)* and *(u)* coefficients of the frequency response (5.44) are given in Table 5.6. Insertion of C_{m0} shifts ω_4 to a frequency much higher than ω_3, so the transfer can be approximated by a three-pole representation. The design criterium to obtain third-order Butterworth poles is then as before:

$$\omega_0^{(m2)} = \frac{g_i}{C_{m2}} = \frac{1}{2}\frac{g_m}{C_{m1}} = \frac{1}{4}\frac{g_o}{C_L + \beta_1 C_{m0}} \tag{5.50}$$

The root loci of the compensation loops then have a similar pattern to those shown in Fig. 5.12, with addition of the non-dominant high frequency pole ω_4. Table 5.7 gives the poles of the OpAmp with a p-n-p Darlington output stage that is introduced in the example in Section 5.1.2. The quiescent current is 500 μA, R_L

Table 5.6. *Coefficients of the characteristic equation of the three-stage OpAmp with Darlington output stage.*
"(nc)"= non-compensated,
"(m1)"= after insertion C_{m1},
"(m2)"= after insertion C_{m2},
"(u)"= after unity-gain feedback.

	$\omega^{(nc)}$	$\omega^{(m1)}$	$\omega^{(m2)}$	$\omega^{(u)}$
ω_0	$\sqrt[3]{\frac{g_i g_m g_o}{C_{m2} C_{m1} C_L}}$	$\sqrt{\frac{g_i g_m}{C_{m2} C_{m1}}}$	$\frac{g_i}{C_{m2}}$	$\frac{2 g_i}{C_{m2}}$
ω_1	$\frac{1}{R_L\left(C_L + \beta_1 C_{m0}\right)}$	$\frac{1}{\beta_1 \beta_2 R_L C_{m1}}$	$\frac{1}{\beta_m \beta_1 \beta_2 R_L C_{m2}}$	$\frac{g_i}{C_{m2}}$
ω_2	$\frac{1}{r_{im} C_{m2}}$	$\frac{1}{r_{im} C_{m2}}$	$\frac{g_m}{C_{m1}}$	$\frac{g_m}{C_{m1}}$
ω_3	$\frac{1}{\beta_2\left(r_{e2} + r_{be1}\right) C_{m1}}$	$\frac{g_o}{C_L + \beta_1 C_{m0}}$	$\frac{g_o}{C_L + \beta_1 C_{m0}}$	$\frac{g_o}{C_L + \beta_1 C_{m0}}$
ω_4	$\frac{1 + \frac{\beta_1 C_{m0}}{C_L}}{r_p C_{be1}}$	$\frac{1 + \frac{\beta_1 C_{m0}}{C_L}}{r_p C_{be1}}$	$\frac{1 + \frac{\beta_1 C_{m0}}{C_L}}{r_p C_{be1}}$	$\frac{1 + \frac{\beta_1 C_{m0}}{C_L}}{r_p C_{be1}}$

Table 5.7. Calculated pole positions for a three-stage OpAmp with a p-n-p Darlington output stage biased at 500 μA, loaded with 100 pF in parallel with 10 kΩ and with a C_{m0} of 20 pF.

	non-compensated $\omega^{(nc)}$	Miller compensated $\omega^{(m1)}$	twice Miller compensated $\omega^{(m2)}$	unity-gain feedback $\omega^{(u)}$
ω_0	1.6 MHz	1.1 MHz	770 kHz	1.6 MHz
ω_1	32 kHz	4 kHz	80 Hz	1.6 MHz, +60°
ω_2	32 kHz	32 kHz	2.2 MHz, +45°	1.6 MHz, 0°
ω_3	380 kHz	3.1 MHz	2.2 MHz, -45°	1.6 MHz, -60°
ω_4	25 MHz	25 MHz	25 MHz	25 MHz
A_0	80 dB	80 dB	80 dB	0 dB

is 10 kΩ, C_L is 100 pF and C_{m0}, C_{m1} and C_{m2} are 20 pF, 10 pF and 5 pF, respectively. If the p-n-p transistors have a transit frequency of 50-MHz and a β of 20, and the β of the n-p-n's is 100, we can calculate all the coefficients of Table 5.6 and solve the characteristic equation of (5.44). The transconductance of the intermediate stage should be 98 μmho, just like g_i in Section 5.1.2, and in this example g_i should be 24 μmho. As mentioned above, insertion of C_{m0} results in a quasi three-pole frequency response, with second-order Butterworth poles in the open-loop gain and third-order poles in the unity-gain transfer.

If the results of this three-stage OpAmp with a p-n-p Darlington output stage are compared with the ones obtained for the single n-p-n output transistor, we find that although both output transistors are biased at the same current of 500 μA, the bandwidth is now almost ten times lower. This is due to the insertion of C_{m0}, which is necessary to assure stability at higher output-current levels. The bandwidth of the two-stage OpAmp with a Darlington output stage, presented in Section 5.1.2, is twice the value obtained here, however, its gain is of course considerably lower.

5.3 MULTI-PATH-DRIVEN MILLER COMPENSATION

The nested Miller compensation, as described in the preceding Section makes it possible to use three or even more amplifier stages in series, and still have an overall frequency response with only one dominating pole frequency. The resulting bandwidth, however, is reduced by a factor of two for every additional nest that is applied. To bypass this disadvantage, an extension to the nested Miller-compensation technique is given in this Section. The multi-path-driven Miller compensation [5] that is introduced here, combines the possibility of a high gain and a high bandwidth. The technique is essentially the same as that presented in Section 3.4, which deals with the multi-path-driven output stage. Then it was applied to combine the good high-frequency behavior of the single CE output transistor with the gain of, for instance, the Darlington output stage. Now we will use it to benefit the frequency behavior of the complete OpAmp.

Fig. 5.15 shows a three-stage OpAmp, similar to that of Fig. 5.10, but extended by a second input stage. This second input stage, Q_{31}, Q_{32}, is directly connected to the output stage, and thus bypasses the intermediate stage. This "feedforward path" therefore only comprises one Miller-compensation loop and has a relatively high bandwidth. The other input stage, Q_{21}, Q_{22}, is as before connected to the

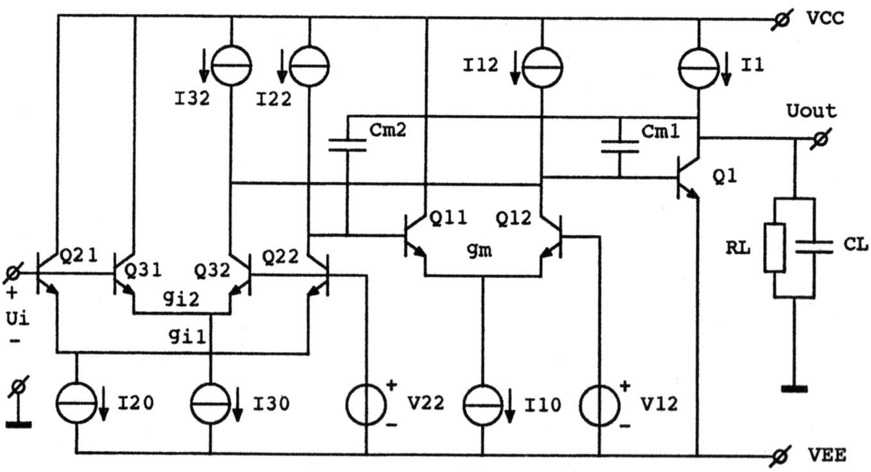

Fig. 5.15. Three-stage OpAmp, compensated by multi-path-driven Miller compensation.

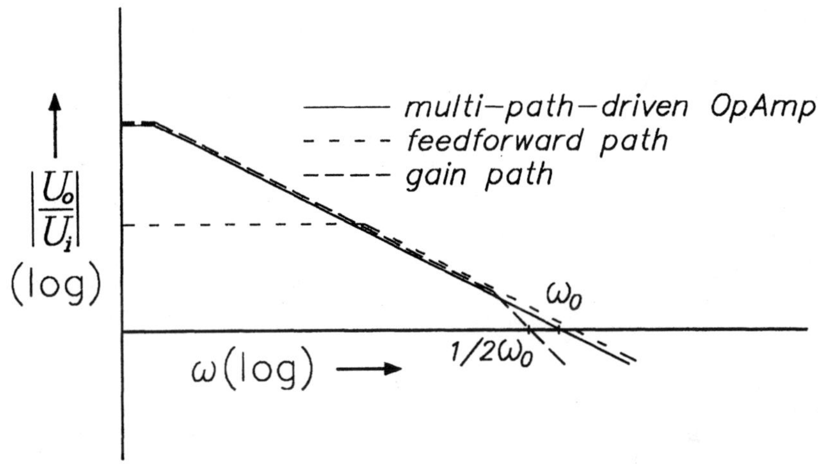

Fig. 5.16. The principle of the multi-path-driven strategy. The overall transfer has the gain of the gain path, and the bandwidth of the feedforward path.

intermediate stage, which in turn supplies the output stage. This path has a bandwidth that is lower, but the gain, on the other hand, is much higher. Fig. 5.16 illustrates the principle of the multi-path-driven strategy. The overall transfer, the continuous line, is composed of the gain part, indicated by the dashed line, and the feedforward part, indicated by the dotted line. This results in a transfer with both a high gain and a high bandwidth.

Although the output stage of the OpAmp shown in Fig. 5.15 consists of just one CE-connected transistor, naturally, the multi-path-driven Miller compensation can also be applied to OpAmps with other output stages. Furthermore, this compensation technique is not limited to the field of low-voltage OpAmps; it can be applied to other OpAmps as well.

The small-signal equivalent of the circuit depicted in Fig. 5.15 is shown in Fig. 5.17. As could be expected, the circuit shown in Fig. 5.17 shows great similarity to the equivalent circuit of the multi-path-driven output stage, shown in Fig. 3.17. If the current equations are solved, the voltage gain of the non-compensated OpAmp can be calculated:

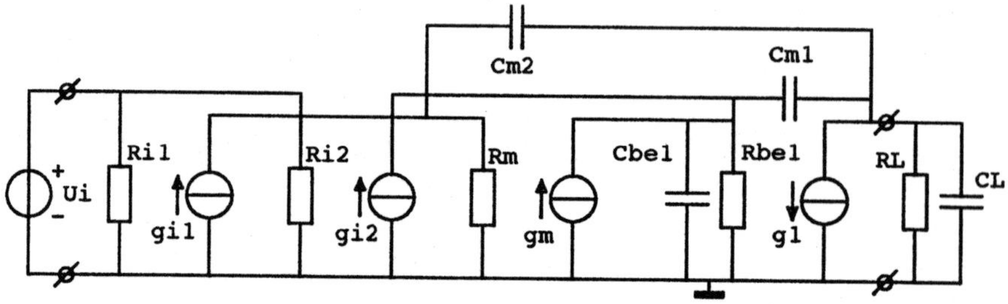

Fig. 5.17. The small-signal equivalent circuit of the OpAmp with multi-path-driven Miller compensation.

$$\frac{U_o}{U_i} = A_0 \frac{1 + \frac{s}{\omega_{11}^{(nc)}}}{1 + \frac{s}{\omega_1^{(nc)}} + \frac{s^2}{\omega_1^{(nc)}\omega_2^{(nc)}} + \frac{s^3}{\omega_1^{(nc)}\omega_2^{(nc)}\omega_3^{(nc)}}} \qquad (5.51)$$

If β_m is the current gain of the intermediate-stage transistors, β_1 that of the output transistor, and g_{i1} is the transconductance of input stage Q_{21}, Q_{22}, the DC voltage gain is:

$$A_0 = -\beta_m \beta_1 g_{i1} R_L \qquad (5.52)$$

The poles of the non-compensated OpAmp, with the Miller capacitors connected to ground instead of to the output, are equal to the poles of the OpAmp without the feedforward path from input stage Q_{31}, Q_{32} to the output stage. The first pole is defined by the input resistance of the intermediate stage and C_{m2}:

$$\omega_1^{(nc)} = \frac{g_m}{\beta_m C_{m2}} = \frac{1}{r_{im} C_{m2}} \qquad (5.53)$$

As in the preceding Section, r_{im} is the input resistance and g_m the transconductance of the intermediate stage. Because the transconductance g_m of the intermediate stage is chosen to be very small, as will be shown later on, this now is the pole with the lowest frequency. The second pole is situated at the output and is determined by the load resistor and capacitor:

$$\omega_2^{(nc)} = \frac{1}{R_L C_L} \tag{5.54}$$

The third pole, at the output of the intermediate stage, is controlled by the input resistance of the output stage and the sum of the diffusion capacity of Q_1 and C_{m1}:

$$\omega_3^{(nc)} = \frac{1}{r_{be1} C_{m1}\left(1 + \frac{C_{be1}}{C_{m1}}\right)} \tag{5.55}$$

The transfer also has a zero, which, once it is matched to one of the poles, will cancel it, and in this way give the circuit a higher bandwidth than it would have without the feedforward path. The zero is:

$$\omega_{11}^{(nc)} = \frac{g_{i1} g_m}{g_{i2} C_{m2}} \tag{5.56}$$

where g_{i2} is the transconductance of the input stage Q_{31}, Q_{32}.

The superscripts *(nc)*, *(mp1)*, *(mp2)* and *(u)* in the Equations indicate whether no compensation, multi-path-driven Miller compensation with C_{m1}, multi-path-driven compensation with C_{m2} or unity-gain feedback is applied to the OpAmp. If the Miller capacitor C_{m1} is connected to the output, the poles at the output and at the output of both the intermediate stage and input stage Q_{31}, Q_{32} are split. The pole at the output of the input stage Q_{21}, Q_{22} and the zero remain unaltered. The coefficients of the voltage-gain transfer (5.51) are replaced by:

$$\omega_1^{(mp1)} = \frac{1}{\beta_1 R_L C_{m1}} \tag{5.57}$$

$$\omega_2^{(mp1)} = \frac{1}{r_{im} C_{m2}} \tag{5.58}$$

$$\omega_3^{(mp2)} = \frac{1}{r_{e1} C_L \left(1 + \frac{C_{be1}}{C_{m1}}\right)} = \frac{g_o}{C_L \left(1 + \frac{C_{be1}}{C_{m1}}\right)} \tag{5.59}$$

and, as stated above, $\omega_{11}^{(mp1)}=\omega_{11}^{(nc)}$. The root locus of this pole splitting at the output is shown at the top of Fig. 5.18, and the movement of the poles in the frequency response of the OpAmp is shown in Fig. 5.19. Except for the zero in the transfer, this is all similar to what happens when nested Miller compensation are applied. If now also C_{m2} is connected to the output, the poles at the output and at the output of the input stage Q_{21}, Q_{22} are moved apart, and the *(mp2)* poles and zero are:

$$\omega_1^{(mp2)} = \frac{1}{\beta_m \beta_1 R_L C_{m2}} \tag{5.60}$$

$$\omega_2^{(mp2)} = \frac{g_m}{C_{m1}} \tag{5.61}$$

The third pole is still determined by the output-stage transconductance and the load capacitor, $\omega_3^{(mp2)}=\omega_3^{(mp1)}$ and also the zero in the transfer is not affected by the compensation, $\omega_{11}^{(mp2)}=\omega_{11}^{(mp1)}$. If the transconductances of the input and intermediate stages, and the Miller capacitors are given adequate values, the second pole $\omega_2^{(mp2)}$, given in (5.61) and the zero $\omega_{11}^{(mp2)}$, given in (5.56) cancel. This happens if:

$$\frac{g_{i1}}{C_{m2}} = \frac{g_{i2}}{C_{m1}} \quad or \quad \frac{g_{i1}}{g_{i2}} = \frac{C_{m2}}{C_{m1}} \tag{5.62}$$

and:

$$\frac{g_m}{C_{m1}} \ll \frac{g_o}{C_L\left(1 + \frac{C_{be1}}{C_{m1}}\right)} \tag{5.63}$$

The current sources I_{20} and I_{30} can be scaled accurately, and because of this the ratio of the transconductances g_{i1} and g_{i2} is fixed. The ratio of the Miller capacitors C_{m2} and C_{m1} is determined by their geometrical dimensions, which are well defined. It is, therefore, very well possible to meet condition (5.62). Further, to satisfy (5.63), the transconductance of the intermediate stage should be sufficiently low, which, for instance, can be realized by inserting resistors in series with the emitters of the intermediate stage transistors Q_{11} and Q_{12}. If (5.62) and (5.63) are met, the transfer is given by:

$$\frac{U_o}{U_i} = A_0 \frac{1 + \frac{s}{\omega_{11}^{(mp2)}}}{\left(1 + \frac{s}{\omega_1^{(mp2)}}\right)\left(1 + \frac{s}{\omega_2^{(mp2)}}\right)\left(1 + \frac{s}{\omega_3^{(mp2)}}\right)} = \frac{A_0}{\left(1 + \frac{s}{\omega_1^{(mp2)}}\right)\left(1 + \frac{s}{\omega_3^{(mp2)}}\right)} \tag{5.64}$$

which is very similar to that of the two-stage OpAmp with just one Miller capacitor, discussed in Section 5.1. The middle part of Fig. 5.18 shows the root locus of the loop gain through C_{m2}. Fig. 5.19 shows the frequency response, after insertion of the Miller capacitors, and under the condition that the pole $\omega_2^{(mp2)}$ and the zero $\omega_{11}^{(mp2)}$ cancel.

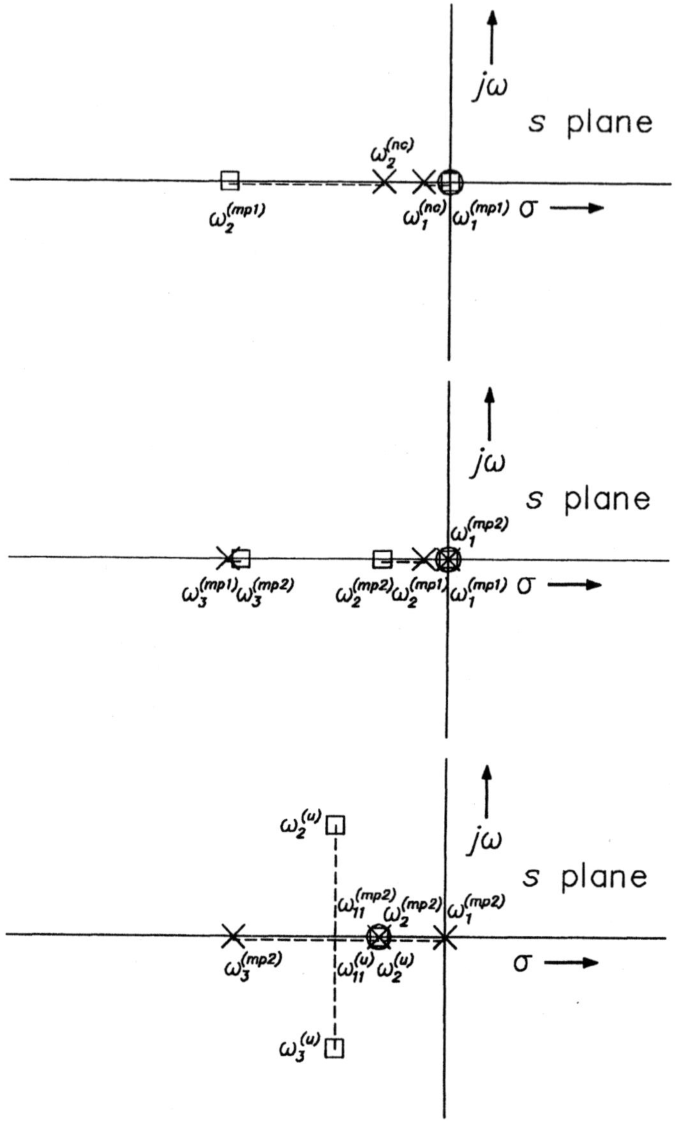

Fig. 5.18. Root loci of: loop gain through C_{m1} and the output stage (top), through C_{m2}, the intermediate and output stage (middle), and the OpAmp, if unity-gain feedback is applied (bottom).

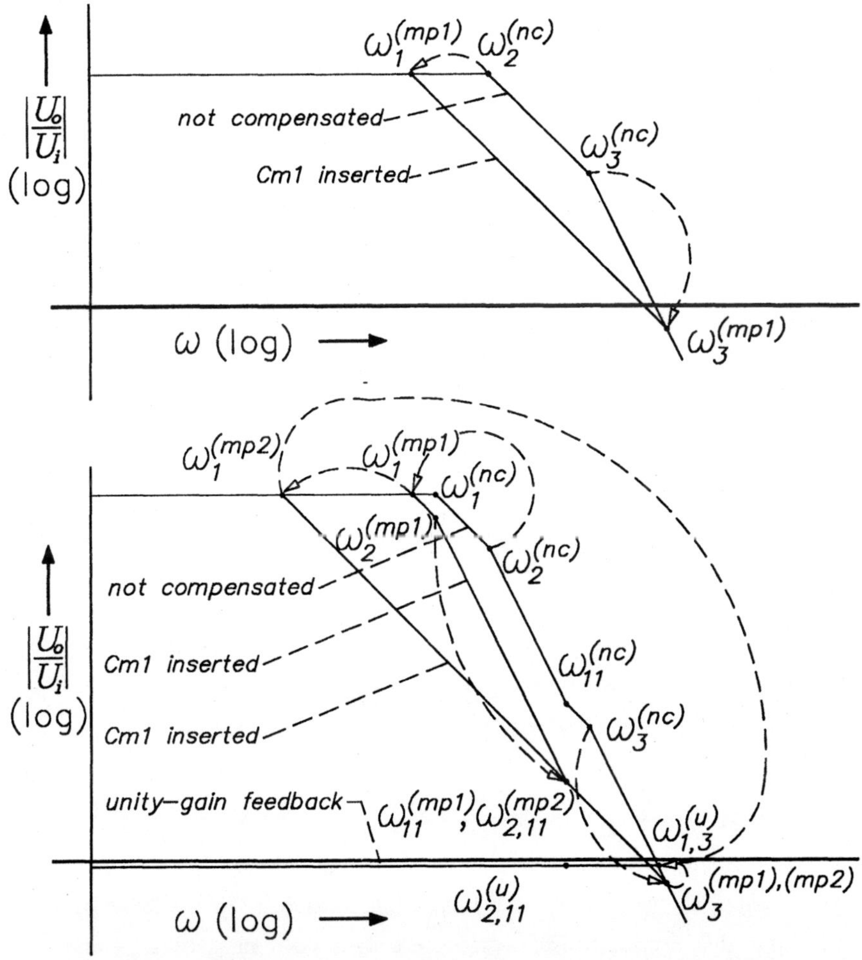

Fig. 5.19. Frequency response of a three-stage OpAmp with multi-path-driven Miller compensation. On top the response of the output and intermediate stage, below that of the complete OpAmp.

To find the appropriate values of the input-stage transconductance in relation to the pole $\omega_3^{(mp2)}$, unity-gain feedback is again applied to the OpAmp, and the resulting characteristic is made equal to the second-order Butterworth transfer (5.11). The voltage-follower function of the OpAmp is:

$$\frac{U_o}{U_i} = \frac{1 + \frac{s}{\omega_{11}^{(u)}}}{1 + s\left(\frac{1}{\omega_1^{(u)}} + \frac{1}{\omega_{11}^{(u)}}\right) + \frac{s^2}{\omega_1^{(u)}\omega_2^{(u)}} + \frac{s^3}{\omega_1^{(u)}\omega_2^{(u)}\omega_3^{(u)}}} \tag{5.65}$$

where $\omega_2^{(u)}$ is equal to $\omega_2^{(mp2)}$ given in (5.61), and the $\omega_3^{(u)}$ is the same as $\omega_3^{(mp2)}$ presented in (5.59). The unity-gain feedback does not affect the zero, so $\omega_{11}^{(u)}$ in (5.65) is the same as $\omega_{11}^{(mp2)}$. Further, $\omega_1^{(u)}$ is:

$$\omega_1^{(u)} = \frac{g_{i1}}{C_{m2}} \tag{5.66}$$

Pole-zero cancellation is again desired, so $\omega_1^{(u)}$ should be much larger than the zero $\omega_{11}^{(u)} = \omega_{11}^{(mp2)}$. In terms of the transconductances of the feedforward input stage Q_{31}, Q_{32} and of the intermediate stage Q_{10}, Q_{11}, this implies:

$$g_{i2} \gg g_m \tag{5.67}$$

This condition is stronger than (5.63), and therefore replaces it. If we consider the feedforward input stage Q_{31}, Q_{32} and the output transistor Q_1, it can be seen in Fig. 5.15 that this amplifier has a feedback path consisting of C_{m2} and the intermediate stage. What (5.67) demands is that this feedback path be suppressed as much as possible.

The unity-gain transfer, if the pole-zero cancelation occurs, is equal to that of a two-stage OpAmp, with one Miller capacitor:

$$\frac{U_o}{U_i} = \frac{1}{1 + \frac{s}{\omega_1^{(u)}} + \frac{s^2}{\omega_1^{(u)}\omega_3^{(u)}}} \tag{5.68}$$

In Section 5.1 it was found that this transfer is equal to the maximally-flat or Butterworth frequency response if $\omega_2^{(u)}=2\omega_1^{(u)}$. The -3-dB bandwidth of the voltage follower is then equal to:

$$\omega_0^{(u)} = \sqrt{2}\,\frac{g_{i2}}{C_{m2}} \tag{5.69}$$

and the open-loop bandwidth, corresponding to (5.64) is:

$$\omega_0^{(mp2)} = \frac{g_{i2}}{C_{m2}} = \frac{1}{2}\,\frac{g_o}{C_L\left(1+\dfrac{C_{be1}}{C_{m1}}\right)} \tag{5.70}$$

The root locus of the OpAmp if unity-gain is applied is shown below in Fig. 5.18, and the unity-gain frequency response is depicted in Fig. 5.19. Table 5.8 shows numerical values of the poles of an OpAmp as in Fig. 5.15, with 500 μA quiescent current through the output transistor Q_1 and loaded with a 100 pF capacitor in parallel with a 10 kΩ resistor. The Miller capacitors C_{m1} and C_{m2} are 10 pF and 5 pF, respectively, and the β of the 5-GHz n-p-n's is 100. In other words, similar conditions as in the example at the end of Section 5.2.1. To satisfy (5.70), g_{i2}, of the feedforward input stage Q_{31}, Q_{32}, has to be 0.9 mmho and, consequently, g_{i1}, the transconductance of the input stages Q_{21}, Q_{22}, is 0.45 mmho. The transconductance of the intermediate stage Q_{11}, Q_{12}, g_m, must be much smaller than g_{i2} and is chosen 90 μmho. Comparing the values in Tables 5.4 and 5.8, we see that the OpAmp with multi-path-driven Miller compensation not only has an open-loop bandwidth that is twice that of the OpAmp with nested Miller compensation, but its gain is also twice as high.

The cancelation of the pole $\omega_2^{(mp2)}$ and the zero $\omega_{11}^{(mp2)}$ depends on the matching of the two input pairs Q_{21}, Q_{22} and Q_{31}, Q_{32}, respectively, and on the matching of the two Miller capacitors C_{m1} and C_{m2}. Since the mismatch can be kept below 1%, the accuracy of the cancelation is very good, and the remaining pole-zero doublet is equally small. In [6] it is shown that the normalized amplitude in the pulse response due to the pole-zero doublet is in first order equal to the doublet spacing divided by the bandwidth. In our example, that is 0.1%. For

Table 5.8. Calculated pole and zero positions for a three-stage OpAmp with multi-path-driven Miller compensation, biased at an output current of 500 μA and loaded with 100 pF in parallel with 10 kΩ.

	non-compensated $\omega^{(nc)}$	multi-path-driven Miller compensated $\omega^{(mp1)}$	twice multi-path-driven Miller compensated $\omega^{(mp2)}$	unity-gain feedback $\omega^{(u)}$
ω_0	21 MHz	15 MHz	15 MHz	21 MHz
ω_1	29 kHz	16 kHz	320 Hz	21 MHz, +45°
ω_2	160 kHz	29 kHz	1.5 MHz	1.5 MHz
ω_{11}	1.5 MHz	1.5 MHz	1.5 MHz	1.5 MHz
ω_3	2.9 MHz	29 MHz	29 MHz	21 MHz, -45°
A_0	93 dB	93 dB	93 dB	0 dB

the 0.1% settling time the doublet thus has no influence. If a higher settling accuracy is demanded, some degradation might occur, however, the influence on the settling time of higher-order poles, that are also present in the conventional Miller-compensated structure, is likely to be at least as large.

The multi-path-driven Miller compensation yields a bandwidth, twice that of the nested Miller compensation, at the cost of an extra input stage and consequently, a somewhat larger chip area and total quiescent-current consumption. Whether or not this is justified depends on the expected applications of the OpAmp. In the case of low-voltage OpAmps with a folded-cascode input stage, as described in Chapter 2, the implementation is rather complex, and the technique is probably only rewarding when a high bandwidth is very essential. A relevant description of the technique applied to a (not low-voltage) OpAmp is [7].

5.4 SLEWING

If large input signals are applied to the OpAmp, the input stage does not operate linearly any more. The current it supplies to the intermediate stage is no longer determined by the small-signal transconductance g_i, but is limited to the maximal value equal to the tail current of the input stage. This constant current charges the Miller capacitor that is used to compensated the intermediate and output stage. The output of the OpAmp then displays a ramp voltage, increasing or decreasing with a constant slope. The change of the output voltage as a function of the time is called the slew rate [2] of the OpAmp. Referring to Fig. 5.1, it can be seen that this slew rate is:

$$S_R = \frac{I_{10}}{2C_m} \tag{5.71}$$

On the other hand, the small-signal transconductance is:

$$g_i = \frac{I_{10}}{4V_T} \tag{5.72}$$

where V_T is the thermal voltage. The bandwidth of the OpAmp expressed in terms of this transconductance is given in (5.14), and combining these relations gives:

$$S_R = 2V_T\omega_0^{(m)} = 4\pi V_T f_0 \tag{5.73}$$

if $f_0 = 2\pi\omega_0^{(m)}$ is the open-loop bandwidth of the OpAmp in Hertz. An OpAmp with a bandwidth of 1 MHz, for example, thus has a slew rate of 0.31 V/μs. This value may be increased by inserting emitter resistors in the transistors of the input stage, but that would increase the offset voltage and the equivalent input noise voltage. In the case of low-voltage OpAmps, the emitter degeneration is normally not applied, and (5.73) holds.

5.5 CONCLUSIONS

To ensure the stability of the OpAmp, frequency compensation is necessary, because there are more than two dominant poles in the transfer from input voltage to output voltage. It was argued that the most effective way to accomplish this compensation is to apply Miller splitting.

The overall bandwidth of the Miller-compensated OpAmp is determined by the transconductance, and therefore the quiescent current of the output stage and by the maximum load capacitor that should be driven. The bandwidth is always at least a factor of two lower than the output pole determined by these quantities. If the output stage introduces complex poles in the frequency response, the stability of the OpAmp can still be guaranteed, but the obtainable bandwidth is very reduced. The use of output stages without these complex poles is therefore emphasized.

If the OpAmp consists of more than two stages its gain increases, and the Miller-compensation technique has to be extended to nested Miller compensation. A disadvantage of the nested Miller compensation is that the bandwidth of the OpAmp is further reduced by a factor two. To by-pass the trade-off between increasing gain and decreasing bandwidth, the multi-path-driven Miller compensation can be used. This technique has the same obtainable bandwidth as the simple Miller compensation, but the gain can be much higher. Depending on the implementation, this method might demand a more complex circuit. Which method should be used depends on the design criteria that are imposed on the OpAmp.

Finally, it was shown that the slew rate of the OpAmp is directly coupled to its open-loop bandwidth.

5.6 REFERENCES

[1] J. E. Solomon, "The Monolithic Op Amp: A Tutorial Study", *IEEE J. Solid-State Circuits*, Vol. SC-9, pp. 314-332, Dec. 1974.

[2] J. H. Huijsing, "Multi-Stage Amplifier with Capacitive Nesting for Frequency Compensation", US. Pat. Appl. Ser. No. 4559502 Dec. 1985.

[3] E. M. Cherry, "A High Quality Audio Power Amplifier", *Monitor, Proceedings of the IREE Aust.*, Jan./Feb. 1978, pp. 1-8.

[4] J. H. Huijsing and D. Linebarger, "Low-Voltage Operational Amplifier with Rail-to-Rail Input and Output Ranges", *IEEE J. Solid-State Circuits*, Vol. SC-20, pp. 1144-1150, Dec. 1985.

[5] J. H. Huijsing and J. Fonderie, "Multi-Stage Amplifier with Capacitive Nesting and Multi-path-Driven Forward Feeding for Frequency Compensation", U.S. Pat. Appl. Ser. No. 654.855, filed February 11, 1991.

[6] B. Y. Kamath, R. G. Meyer and P. R. Gray, "Relationship Between Frequency Response and settling time of Operational Amplifiers", *IEEE J. Solid-State Circuits*, Vol. SC-9, Dec 1974, pp. 347-352.

[7] R. G. H. Eschauzier, L. P. T. Kerklaan and J. H. Huijsing, "A 100MHz 100dB Operational Amplifier with Multi-path Nested Miller Compensation Structure", in *ISSCC Dig. Tech. Papers*, 1992, pp. 196-197.

CHAPTER 6
REALIZATIONS

This Chapter is dedicated to the discussion of complete OpAmp designs compiled of the circuit parts that have been described in the previous Chapters. Section 6.1 describes the design of the commercially available 2-V OpAmp NE5234. This OpAmp was developed for the Signetics Company as follow-up to the low-voltage OpAmp NE5230 [1]. In Section 6.2 the 1-V OpAmp IS549, processed at the Delft Institute of MicroElectronics and Sub-micron technology (DIMES). To compare the performance of the output stages discussed in Chapter 3, four OpAmps with similar input and intermediate stages, but with four different output stages were integrated. These four OpAmps are discussed in Section 6.3. The OpAmp that is discussed in Section 6.4 is a 1-V, 10-MHz OpAmp, designed as building block for an analog cell library of the Signetics Company, rather than as stand-alone device.

6.1 2-V OPAMP NE5234

As the successor to Signetics' low-voltage OpAmp NE5230, a redesign, the NE5234, was realized using their HS3 process. The HS3 process is a 7-V oxide-isolated process with 10-GHz n-p-n, and 160-MHz p-n-p transistors.

The output stage of the NE5234, shown in Fig. 6.1, comprises an n-p-n Darlington pair Q_{210} and Q_{220}, and an extended p-n-p Darlington Q_{110}, Q_{120} and Q_{130}. To be able to push output currents up to 10 mA, the output transistor Q_{110} is a 32× lateral p-n-p, and its driver Q_{120} is a 8× vertical p-n-p. Transistor Q_{130} precedes Q_{120} to further boost the gain, and to restore the voltage level at the input

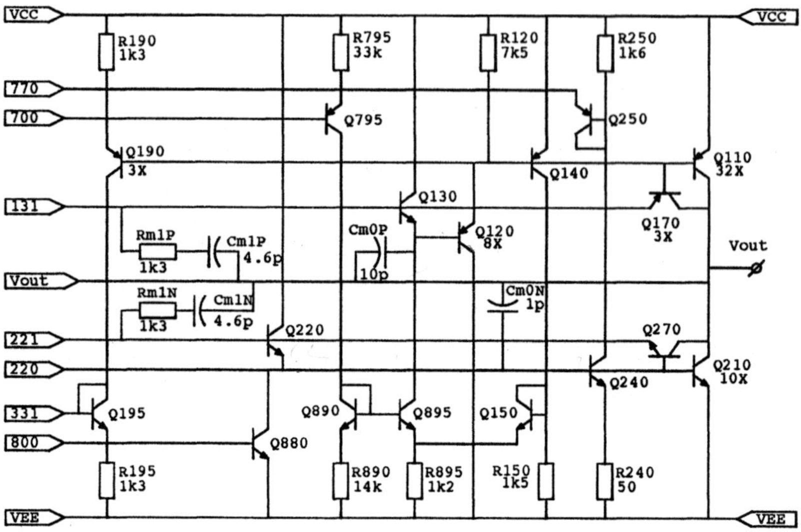

Fig. 6.1. Complementary output stage of the NE5234 OpAmp.

of the output stage. The n-p-n driver Q_{220} is biased at a quiescent current of 10 μA by current source Q_{880}. This current source is connected to the n-p-n biasing rail, labelled "800". From R_{120} a quiescent current of 100 μA flows through the p-n-p driver Q_{120}. The second driver of the p-n-p output stage, Q_{130}, is biased at 45 μA by Q_{890} and Q_{895} and by Q_{795}, connected to the p-n-p biasing rail "700". The currents through the emitter followers of the p-n-p output stage had to be chosen relatively large, to limit the deterioration of the frequency response.

The sensing of the output current for the benefit of the class-AB current-control loop is done on the base of Q_{210}, and on the base of the diode Q_{195}, which forms with Q_{190} an image of the p-n-p output current.

The saturation of the output transistors Q_{110} and Q_{210} is sensed with Q_{170} and Q_{270}, respectively, as has been described in Section 4.3. If the output current tends to exceed safe values, it is limited to +/-15 mA, to prevent destruction of the output stage. Transistor Q_{140} senses the p-n-p output current, and after the threshold, built in with R_{150} and Q_{150}, is exceeded, current source Q_{895} is deactivated. This in turn removes the base current of Q_{120}, which is thereby shut off. The n-p-n output current is sensed with Q_{240}, and when the voltage over R_{250}

allows the diode Q_{250} to conduct, the current from the intermediate-stage current source Q_{770} is removed, and Q_{220} loses its input current.

Fig. 6.2 shows the intermediate stage and class-AB current control. The class-AB current control is similar to that shown in Fig. 4.7. Transistors Q_{330} and Q_{340} compare the currents through Q_{110} and Q_{210} by measuring the voltages on the nodes labelled "331" and "220", respectively, and with the feedback amplifier Q_{310}, Q_{320}, the smallest of these currents, is matched to the current through the diodes Q_{920} and Q_{930}. The quiescent current through Q_{110} and Q_{210} is regulated at 310 μA. If Q_{920} is matched to Q_{330} and Q_{340}, and Q_{930} is of the same type as the n-p-n output transistor Q_{210}, the class-AB current control is compensated for temperature variations as well as for irregularities in the processing parameters.

The common-mode values of the input voltages of the output stage, labelled "131" and "221" in Figs. 6.1 and 6.2, vary with the supply voltage, and therefore the intermediate-stage transistors and the class-AB current control suffer from the Early effect. As discussed in Chapter 4, the quiescent current through the output transistors will therefore be sensitive to variations in the supply voltage. To compensate this, the reference voltage on the base of Q_{310}, determining the quiescent current, is also exposed to the Early effect; the part of the current through the diodes Q_{920} and Q_{930} coming from Q_{840}, Q_{780}, Q_{785}, Q_{850} and Q_{860} also varies with the supply voltage.

The current through Q_{310} and Q_{320} is 9 μA, equal to that through each of the intermediate stage transistors Q_{410}-Q_{440}. To ensure the stability of the class-AB control loop, the transconductance of the feedback amplifier Q_{310}, Q_{320} is lowered by inserting emitter resistors R_{310} and R_{320}.

The intermediate stage of the NE5234 is also shown in Fig. 6.2. Transistors Q_{410}-Q_{440} form the intermediate stage as characterized in Section 4.1, and Q_{450} and Q_{460} are emitter followers to boost the current gain. The tail-current source Q_{830} steers 9 μA through each of the transistors Q_{410} through Q_{440}, and the current through the emitter followers Q_{450} and Q_{460} is 4 μA each. A biasing loop through the summing-circuit transistors Q_{590} and Q_{595}, shown in Fig. 6.3, further comprises the Schottky diode D_{820} and the intermediate-stage transistors Q_{410}-Q_{460}. The small voltage drop over this Schottky diode fixes the voltage on the collector of Q_{830} at 460 mV above the negative supply rail V_{EE}. The common-mode voltage

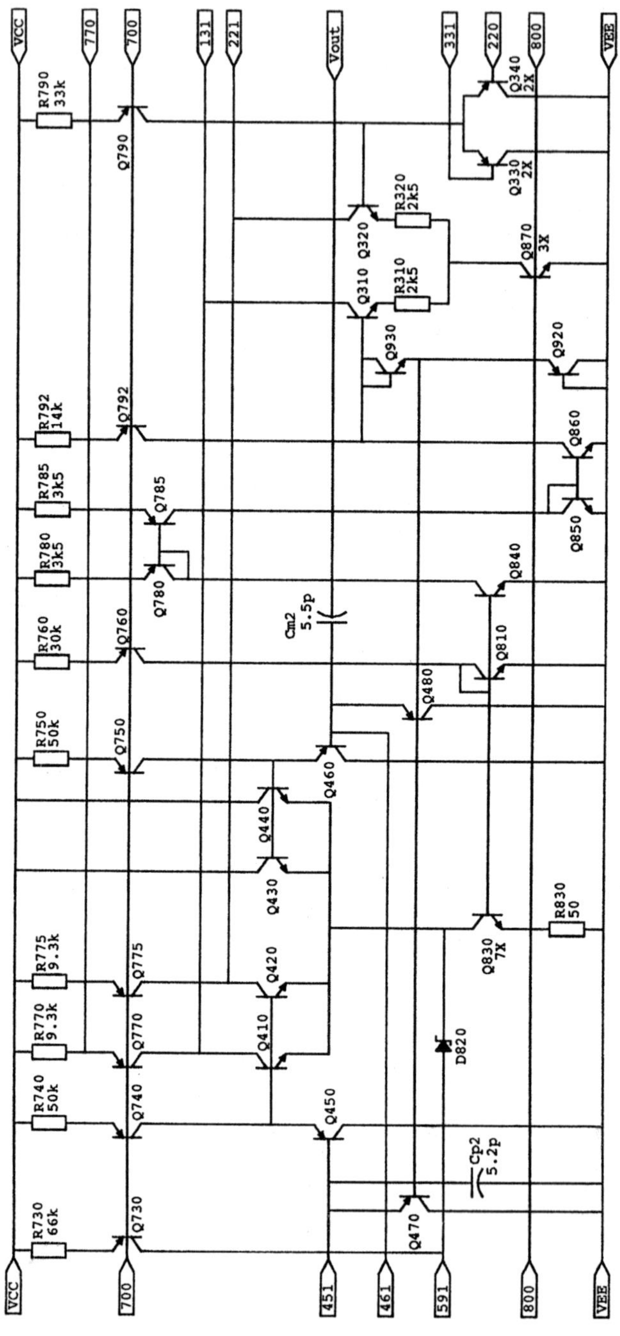

Fig. 6.2. Intermediate stage and class-AB current control of the NE5234 OpAmp.

on the bases of Q_{450} and Q_{460} is 520 mV above V_{EE}, and summing-circuit transistors Q_{590} and Q_{595} now have a collector-emitter voltage of 350 mV. Transistors Q_{470} and Q_{480} clamp the voltage on the bases of Q_{450} and Q_{460}, respectively, in order to prevent saturation of current sources Q_{740} and Q_{750} if the inputs of the OpAmp are driven hard.

Fig. 6.3. shows the input stage of the NE5234 OpAmp. The complementary input pairs Q_{510}-Q_{520} and Q_{530}-Q_{540} are identical to the input stage described in Section 2.2. The input stage has a constant transconductance over the full common-mode input range, that extends from the positive to the negative supply rail. The operation of the reversal-protection diodes Q_{515}, Q_{525}, Q_{535} and Q_{545} was described in Section 2.4. The functioning of the summing circuit Q_{580}-Q_{595} was also addressed in Chapter 2. The summing circuit now has a differential output, and the transistors Q_{590} and Q_{595} are biased with a common-mode feedback loop consisting of these transistors and the intermediate stage. The current of tail-current source Q_{710}, as well as that through Q_{580} and Q_{585}, is 5.5 μA. Current source Q_{720} supplies 3 μA to Q_{910} and R_{910}, giving a reference voltage at the base

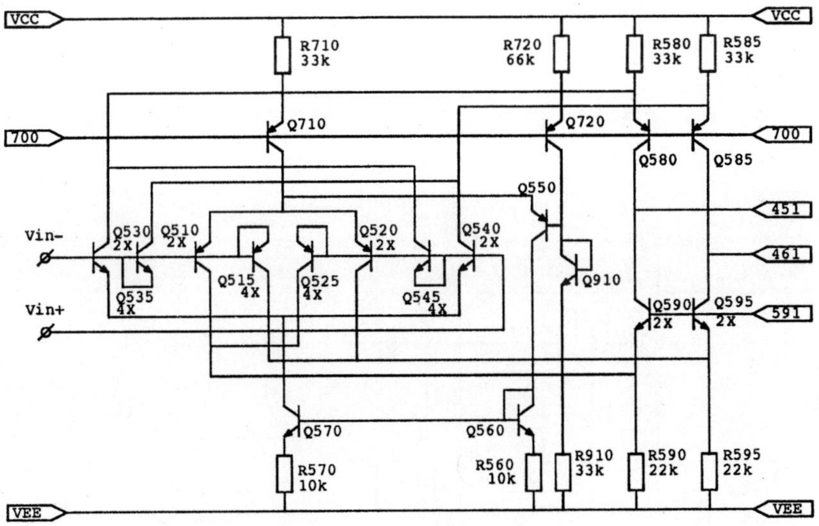

Fig. 6.3. Input stage of the NE5234 OpAmp.

of current switch Q_{550} of 0.8 V above the negative supply rail. All of these current sources are connected to the p-n-p biasing rail of the PTAT-current source, indicated in Fig. 6.3 by the label "700".

The frequency compensation of the output stage is realized with Miller capacitors C_{m1N} and C_{m1P}. In series with these capacitors, the small resistors R_{m1P} and R_{m1N} are inserted to increase the phase margin of the frequency response. At large output currents, the poles of the output stage tend to become complex, as was elaborated in Chapter 5. To increase the damping ratio of these complex poles, the additional Miller capacitors C_{m0P} and C_{m0N} are also inserted. The complete OpAmp is compensated by C_{m2} and C_{p2}, shown in Fig. 6.2. They give the NE5234 a open-loop bandwidth of 2.5 MHz, with a phase margin of 58°, when it is loaded with a 10 kΩ resisotr and a 100 pF capacitor in parallel.

The PTAT reference-current generator is shown in Fig. 6.4. The actual PTAT-current source consists of Q_{10}, Q_{20}, Q_{30}, Q_{40}, and R_{40} and R_{41}. The insertion of the cascodes Q_{50} and Q_{85} eliminates the effect of supply-voltage variations, and the diodes Q_{65} and Q_{90} give the p-n-p and n-p-n output rails "700" and "800", respectively, a low output impedance. The start-up resistor R_{70} makes sure that the currents through the circuit does not remain zero if the power is switched on.

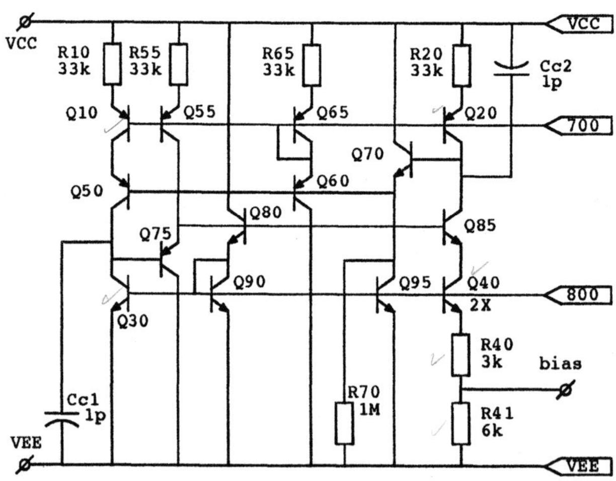

Fig. 6.4. Bias-current generator of the NE5234 OpAmp.

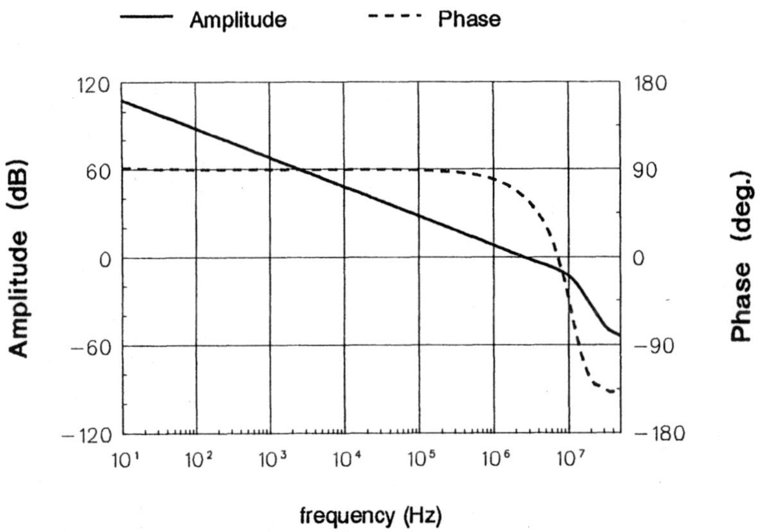

Fig. 6.5. Open-loop frequency response of the NE5234. V_{sup}= 2 V, T= 27 °C, R_L= 10 kΩ, C_L= 100 pF.

In the standard version of the NE5234, the output pin labelled "bias" is connected to the negative supply rail, and the second PTAT resistor R_{41} is short-circuited. The total PTAT resistance is now 3 kΩ, and the current through the branches of the PTAT circuit is 6 μA. If, at assembly, the by-pass is not applied, the PTAT resistance increases to 9 kΩ, and the current through Q_{40} and the other transistors drops to 2 μA. This implies that all current sources of the OpAmp are then also biased at a lower current, and consequently, the power consumption of the circuit is reduced. However, the transconductances of the output stage, intermediate stage and input stage are more or less directly dependent on the value of the PTAT current, and therefore they are also reduced, each in an equal amount relative to the other. The bandwidth in this "low-bias" mode of the OpAmp is therefore proportionally reduced. The total current consumption of the NE5234 in high-bias mode is 700 μA, and this drops to 200 μA in the low-bias mode. The bandwidth is then reduced to 700 kHz.

Table 6.1 gives a summary of the specifications of the NE5234, and in Fig. 6.5 the frequency response is shown. The specs show that the NE5234 is a robust low-voltage OpAmp, that can be operated at supply voltages down to 2 V, in a

Table 6.1. Specifications of the NE5234. V_{sup}= 2 V, T= 27 °C, R_L= 10 kΩ, C_L= 100 pF.

symbol	parameter	low bias	high bias	unit
T_{OP}	operation temperature range	-55/125	-55/125	°C
V_{CC}	supply voltage range	2-7.5	2-7.5	V
I_{CC}	supply current	200	690	μA
V_{OS}	offset voltage	0.4	0.4	mV
I_B	input bias current	32	100	nA
V_{CM}	common-mode input range	V_{EE}-0.25/ V_{CC}+0.25	V_{EE}-0.25/ V_{CC}+0.25	V
CMRR	common-mode rejection ratio	85	85	dB
I_L	peak-load current	±5	±15	mA
A_{VOL}	open-loop voltage gain		120	dB
V_{OUT}	output-voltage swing	V_{EE}+0.1/ V_{CC}-0.1	V_{EE}+0.1/ V_{CC}-0.1	V
S_R	slew rate	0.18	0.69	V/μs
BW	open-loop bandwidth	0.7	2.5	MHz
θ_M	phase margin	61	59	°
NOISE	input referred noise voltage at 1kHz	49	19	nV/√Hz

temperature range from -55 to -125 °C. The input signals may have a common-mode voltage value anywhere in the supply voltage range, and also the output voltage is allowed to swing from rail to rail. An output current of +/-15 mA can be pushed or pulled from the load, and the bandwidth is 2.5 MHz. As can also be seen from Table 6.1, the power consumption can be reduced if the OpAmp is used in the low-bias mode. The specifications such as bandwidth, slew rate and output current capability are then reduced by a factor of three.

6.2 1-V OPAMP IS549

The first 1-V OpAmp to be discussed in this Chapter is the IS549 [2]. It is a truly 1-V OpAmp, with a rail-to-rail input-voltage range that involves a dynamic level-shift, and also a rail-to-rail output-voltage range. The chip is designed in a conventional 18-V process, with n-p-n transistors with an f_T of 300 MHz, and p-n-p's with an f_T of 3 MHz. The processing was carried out by the Delft Institute of MicroElectronics and Sub-micron Technology. The low p-n-p transit frequency limits the obtainable bandwidth, which for this device is chosen at 450 kHz.

The output stage of the IS549 is shown in Fig. 6.6. It consist of two complementary Widlar output stages, both of which are similar to the one described in Section 3.3. The p-n-p half comprises the output transistor Q_{110}, and the two folded emitter followers Q_{120} and Q_{130}. To be able to handle output currents up to and over 10 mA, output transistor Q_{110} is made a 28× lateral p-n-p. The base-emitter voltage of Q_{120} always has to be smaller than that of Q_{110}, and therefore it is also a large transistor. Because Q_{120} is an n-p-n, this does not induce a substantial chip-area increase. To keep the minimum supply voltage at 1 V, the base-emitter voltage of Q_{130} has to be as small as possible, thus it too has to be a large (28×) lateral p-n-p. The boost circuit that provides Q_{110} with sufficient base current at large output currents, consists of Q_{140}, Q_{145} and Q_{147}. The resistor R_{145} in series with the emitter of Q_{145} is inserted to further amplify the current through the current mirror Q_{145}, Q_{147}. The current source Q_{890}, Q_{880}, Q_{795} gives Q_{120} a bias current of 10 μA, which prevents instability at small output currents, as was explained in Section 3.3.

The n-p-n half of the output stage is completely complementary to the p-n-p half, with output transistor Q_{210}, emitter followers Q_{220} and Q_{230}, boost circuit Q_{240}, Q_{245} and Q_{247}, and current source Q_{797}. This half, too, embodies large

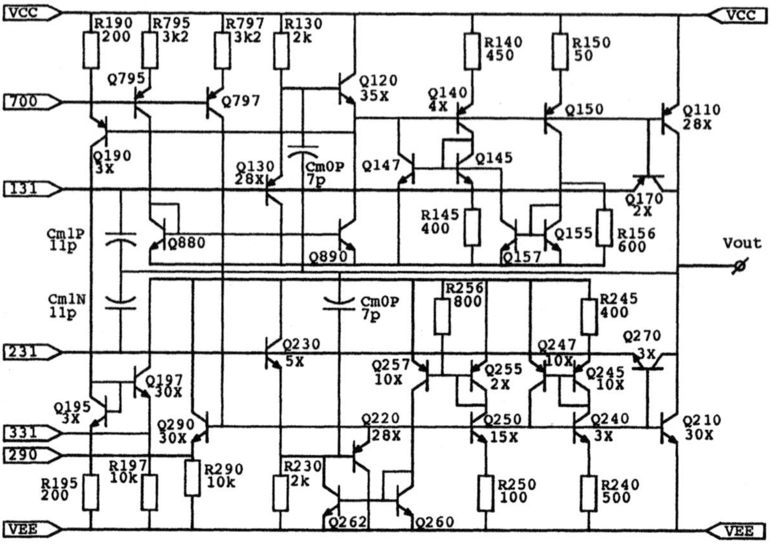

Fig. 6.6. Output stage of the IS549 OpAmp.

transistors, which is necessary to keep the minimum supply voltage as low as possible.

The currents through the output transistors are controlled by the class-AB loop through Q_{290} and the identical Q_{197}. The base of the n-p-n output transistor Q_{210} is directly accessed by Q_{290}, and Q_{197} is connected to the base of p-n-p output transistor Q_{110} through Q_{195} and Q_{190}. The quiescent current through Q_{110} and Q_{210} is 210 μA, and the total current consumption of the output stage is 560 μA.

Transistors Q_{170} and Q_{270} protect the output transistors against heavy saturation, and excessive output currents are limited by Q_{150}, Q_{155} and Q_{157}, and Q_{250}, Q_{255}, Q_{257}, Q_{260} and Q_{262}, for the p-n-p and n-p-n half, respectively. The p-n-p limiter cancels the current boosting of Q_{140}-Q_{147}, and the n-p-n limiter simply takes away the input current of Q_{220}. Thus, the output current is limited at +/-15 mA.

Fig. 6.7 depicts the intermediate stage and class-AB current control of the IS549. The class-AB control is almost identical to that described in the previous Section. The decision pair Q_{330}, Q_{340} decides which of the currents through the output transistors is the smallest. Feedback amplifier Q_{310}, Q_{320} controls the output currents by steering a correction signal if necessary, and the current sources Q_{840},

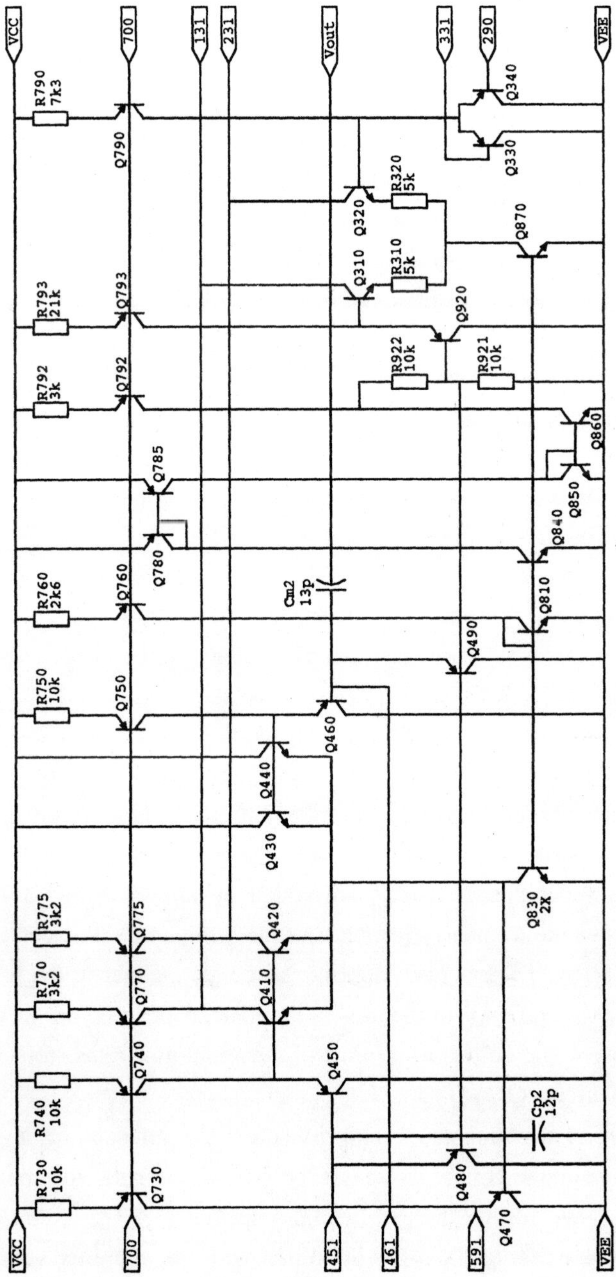

Fig. 6.7. Intermediate stage and class-AB current control of the IS549 OpAmp.

Q_{780}, Q_{785}, Q_{850}, Q_{860}, and Q_{792}, supplying the voltage over R_{921}, again compensate for the Early effect on the intermediate stage and class-AB control transistors. The current source Q_{870} steers 10 μA through Q_{310} and Q_{320}, and through Q_{330} and Q_{340} flows 3 μA each. The reference-voltage source of the feedback amplifier Q_{310}, Q_{320} now consists is lower than that of the NE5234, because the IS549 is a 1-V OpAmp. It consists of Q_{920}, which is matched to Q_{330} and Q_{340}, and of the resistor R_{921}, which is identical to R_{197} and R_{290} in Fig. 6.6. Resistor R_{922} is inserted as a level shift, to give Q_{860} a little more headroom.

The intermediate stage is a duplicate of that of the NE5234. The differential stage Q_{410}-Q_{440} is cascaded with emitter followers Q_{450} and Q_{460}, and the clamps Q_{480} and Q_{490} are inserted to prevent saturation of current sources Q_{740} and Q_{750} if the input of the OpAmp is driven hard. The tail-current source Q_{830} biases the differential stage at 20 μA, and through the emitter followers Q_{450} and Q_{460} flows 5 μA. The common-mode feedback loop through the intermediate stage, through Q_{470} and through the transistors of the summing circuit of the input stage, fix the voltage at the collector of Q_{830} at 150 mV above the negative supply V_{EE}, and the voltage at the bases of Q_{450} and Q_{460} at 160 mV above V_{EE}.

Fig. 6.8 shows the 1-V rail-to-rail complementary input stage with dynamic level-shift of the IS549. This is an implementation of the input stage that was described in Section 2.3. The p-n-p input pair Q_{510}, Q_{520}, together with reversal-protection diodes Q_{515} and Q_{525} are connected to the inputs of the OpAmp through the level-shift resistors R_{511} and R_{521}. Similarly, the n-p-n input pair Q_{530}, Q_{540}, together with Q_{535} and Q_{545} are connected to the inputs through R_{531} and R_{541}. The feedforward capacitors C_{511}-C_{541} are added in parallel with the level-shift resistors to improve the frequency response of the input stage. The current switch Q_{550} directs the 4 μA tail current of current source Q_{710} either to the p-n-p input pair, or through Q_{560} and Q_{570} to the n-p-n input pair, depending on the common-mode input voltage. The collector currents of the input-pair transistors are added by the summing circuit Q_{580}-Q_{595}.

The dynamic level-shift generator comprises the differential amplifier Q_{610}, Q_{620} that compares the voltage at the emitters of the n-p-n input pair with the reference voltage over R_{621}, and further the differential amplifier Q_{630}, Q_{640} that compares the voltage at the emitters of the p-n-p input pair with the reference voltage at the

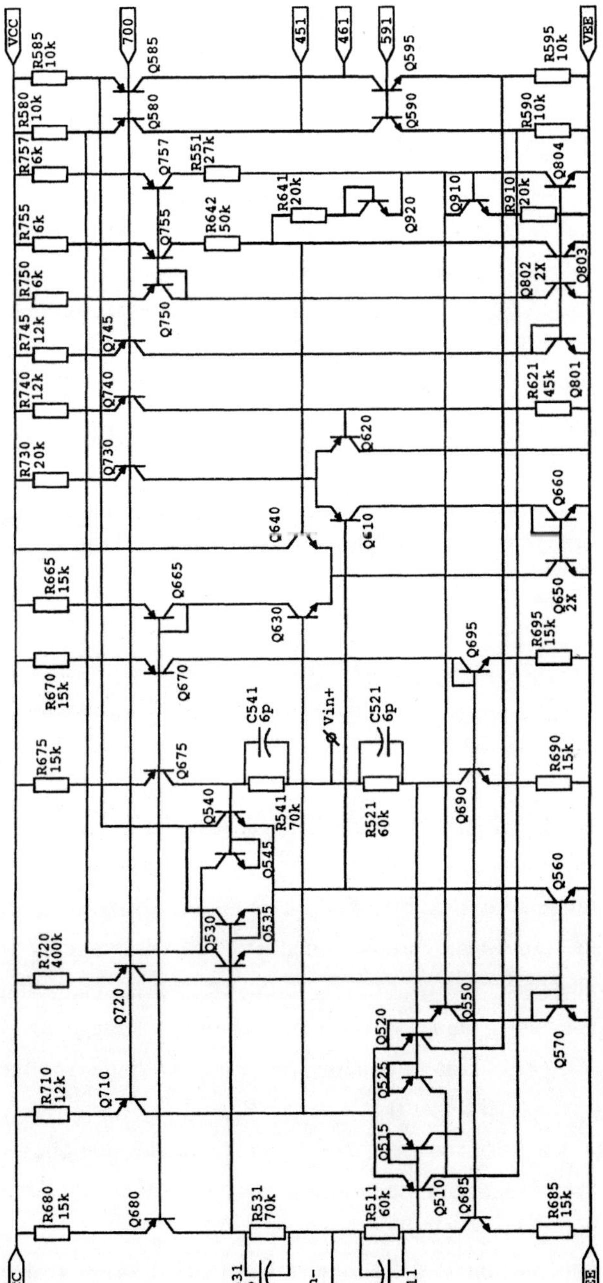

Fig. 6.8. Input stage and dynamic level shift of the IS549 OpAmp.

base of Q_{640}. Depending on the common-mode input voltage and on the supply voltage, this dynamic level-shift generator directs just enough of the current of current sources Q_{670}, Q_{675}, Q_{680}, and Q_{685}, Q_{690}, Q_{695} to the level-shift resistors, and thus enables either the p-n-p or the n-p-n input pair to operate.

At a 1-V supply, the reference voltages at the bases of Q_{550} and Q_{640} have values of 0.75 V and 0.25 V below the positive supply rail V_{CC}, and the voltage over R_{621} is 0.2 V. This is in agreement with the theory of Section 2.3. Now suppose that the supply voltage is much higher than the minimum value of 1 V. The voltage at the base of Q_{640} is then almost equal to V_{CC}. If the common-mode input voltage is equal to V_{EE}, the voltage at the base of Q_{630} is almost equal to V_{EE} too, and therefore Q_{630} is reverse biased with almost the total supply voltage. Since n-p-n transistors will zener if they are reverse biased at voltages more than about 6 V, the circuit is implemented in such a way that when the supply voltage increases, and the level shift is deactivated, the base voltage of Q_{640} is not referenced to the positive, but to the negative supply rail. At low supply voltages, the diodes Q_{910} and Q_{920} are not able to conduct, and therefore current source Q_{755}, that tries to source twice the 4 μA current of Q_{803}, saturates. The voltage at the base of Q_{640} is then determined by the small saturation voltage of Q_{755} and by the voltage over R_{642}. If the supply voltage increases, diode Q_{920} eventually starts to conduct, and Q_{755} comes out of saturation. The voltage at the base of Q_{640} is then about 1.5 V above V_{EE}.

The voltage at the base of current switch Q_{550} is realized in a similar way. If this voltage is referenced to V_{EE} instead of to V_{CC}, the larger part of the common-mode input-voltage range is covered by the n-p-n input pair, which performs better in so far as offset and input-bias current is concerned. At low supply voltages, Q_{757} saturates, and the base voltage of Q_{550} is determined by the voltage over diode Q_{925} and resistor R_{551}, and by the small saturation voltage of Q_{757}. At higher supply voltages, Q_{910} starts to conduct, and Q_{757} no longer saturates. The base voltage of Q_{550} is now fixed at 0.7 V above V_{EE}.

The PTAT-current generator shown in Fig. 6.9 is identical to the one described in Section 4.5. The current through the branches Q_{10}-Q_{30} and Q_{20}-Q_{40} is 4.8 μA, and the start-up current through Q_{70} is 0.3 μA.

The frequency compensation of the output stage of the IS549 is realized with

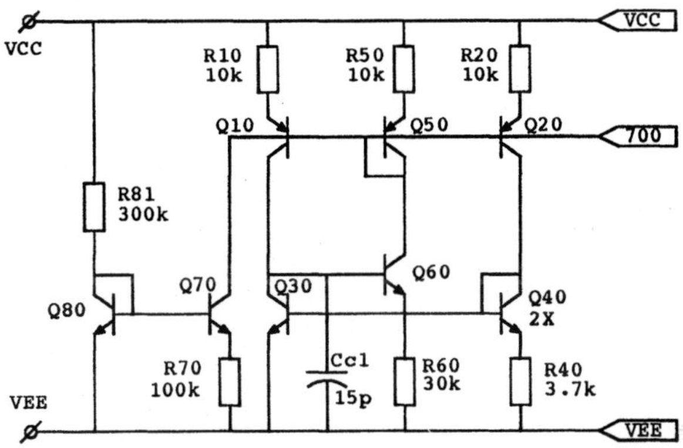

Fig. 6.9. PTAT-current generator of the IS549 OpAmp.

Miller capacitors C_{m1P} and C_{m1N}, of 11 pF each. To restrain complex poles at large output currents, C_{m0P} C_{m0N} of 7 pF each are also inserted. The complete OpAmp is compensated with C_{m2}, which is 13 pF, and C_{p2} of 12 pF. The latter is smaller, because of the parasitic capacitance that already exists at the output of the input-stage summing circuit. The overall bandwidth of the OpAmp is 450 kHz, at a 53° phase margin. The measured frequency response of the IS549 is shown in Fig. 6.10.

The offset voltage of a typical device at a 1-V supply voltage is shown in Fig. 6.11. The step in the middle of the common-mode input voltage represents the range where the level shift is most active. The measured offset voltage and *CMRR* are larger than could be expected from the theory presented in Chapter 2. This is the result of poor process performance, which was, unfortunately, much worse than expected; at the time of integration, the chip-processing work-shop was transformed into the Delft Institute of Microelectronics and Sub-micron technology, and had to move to a new location. Fig. 6.11 also shows that the common-mode input range exceeds both supply rails for about 300 mV, as was predicted in Section 2.4.

Table 6.2 summarizes the specifications of the IS549. The temperature range from 0 °C to +125 °C is associated with the minimum supply voltage of 1 V.

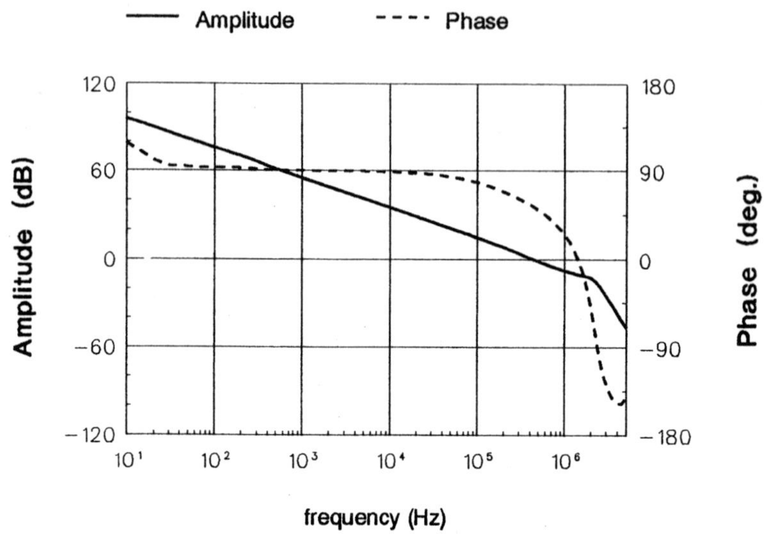

Fig. 6.10. Measured open-loop frequency response of the IS549.
V_{sup}= 1 V, T= 27 °C, R_L= 10 kΩ, C_L= 100 pF.

Extending the temperature range down to -55 °C, takes a minimum supply voltage of 1.2 V. The total supply current of 750 μA is constant within 20 percent over the supply-voltage range from 1 to 10 V. With respect to the worst-case *CMRR*, it can be seen from Fig. 6.11 that the 38-dB value of is in the middle of the common-mode range; the *CMRR* greatly improves when the common-mode input voltage is near either one of the supply rails, or when the supply voltage is increased. A micrograph of the OpAmp is shown in Fig. 6.12. On the left-hand side the level-shift resistors and feedforward capacitors can be distinguished. To the right of the level-shift resistors are the input transistors, which have a quad layout to minimize the offset voltages, as was explained in Chapter 2. On the right-hand side of the chip the 28x lateral output p-n-p and the other large transistors of the output stage can be seen. The total chip measures 2.5 × 5.5 mm^2 (98 × 216 mil^2), which is not suprisingly large, since one p-n-p transistor in this conventional process already requires 100 × 100 μm^2.

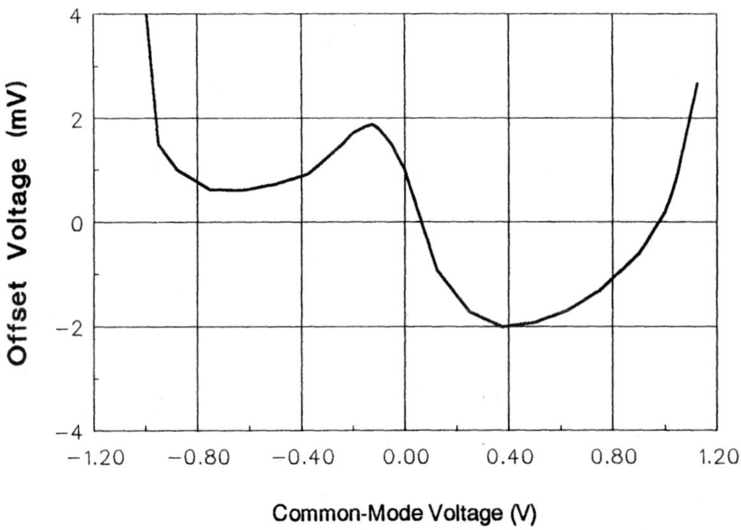

Fig. 6.11. *Offset voltage of an arbitrary device as function of the common-mode input voltage at a supply voltage of 1 V.*

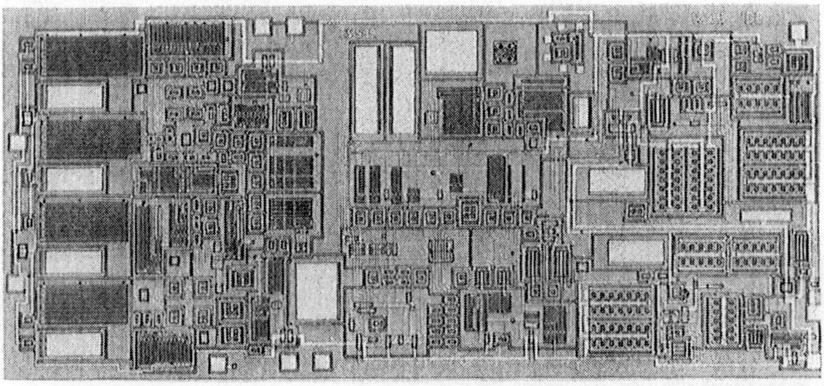

Fig. 6.12. *Micrograph of the IS549. The chip measures 2.5 × 5.5 mm² (98 × 216 mil²).*

Table 6.2. Specifications of the IS549. V_{sup}= *1 V, T= 27 °C,* R_L= *10 kΩ,* C_L= *100 pF.*

symbol	parameter	value	unit
T_{OP}	operation temperature range	-55/125	°C
V_{CC}	supply voltage range	1-10	V
I_{CC}	supply current	750	μA
V_{OS}	offset voltage	+2.0/-2.0	mV
I_B	input bias current	20-175	nA
V_{CM}	common-mode input range	V_{EE}-0.3/ V_{CC}+0.3	V
CMRR	common-mode rejection ratio	100-35	dB
I_L	peak-load current	±10	mA
A_{VOL}	open-loop voltage gain	100	dB
V_{OUT}	output-voltage swing	V_{EE}+0.1/ V_{CC}-0.05	V
S_R	slew rate	0.17	V/μs
BW	open-loop bandwidth	0.45	MHz
θ_M	phase margin	53	°
NOISE	input referred noise voltage at 1kHz	75	nV/√Hz

6.3 OPAMPS U2010 AND U2011

To be able to compare the performance of the Darlington, Widlar, multi-path-driven, and parallel-feedback-compensated output stages, all of which have been discussed in Chapter 3, four OpAmps have been integrated, each with one of these output stages. For practical reasons, two chips have been made, the U2010, comprising the OpAmps with Darlington and multi-path-driven output stages, and the U2011, with the OpAmps with Widlar and parallel-feedback-compensated output stages. All designs were part of a multi-project chip, in a complementary 12-V BiMos process, with 5-GHz n-p-n and 3-GHz p-n-p transistors, processed by Philips Components, Nijmegen.

The Darlington, Widlar, multi-path-driven, and parallel-feedback-compensated output stages will be discussed in Sections 6.3.1 through 6.3.4, respectively. To be able to make a fair comparison, all four OpAmps have the same input and intermediate stages. These common circuit parts are the topic of Section 6.3.5.

The output transistors of the four output stages are biased at an equal quiescent current. Thus, the maximal obtainable bandwidth, determined by the quiescent current and the load capacitor, is equal for all OpAmps. In accordance with the theory set out in Chapters 3 and 5, the transistors in the signal path that precede the output transistors lower the bandwidth, and to limit this reduction these preceding drivers have to be biased at rather large quiescent currents. A measure for the quality of the output stage is therefore not only its bandwidth, but also the total current consumed by the output stage, at zero output current. Since operation at low supply voltages is inherently important to low-voltage OpAmps, this too determines the suitability of the output stages. In Section 6.3.6, dealing with the measurement results on the four OpAmps, a Figure of merit is therefore introduced to express the dependency of the quality of the output stage on these parameters.

6.3.1 Darlington Output Stage

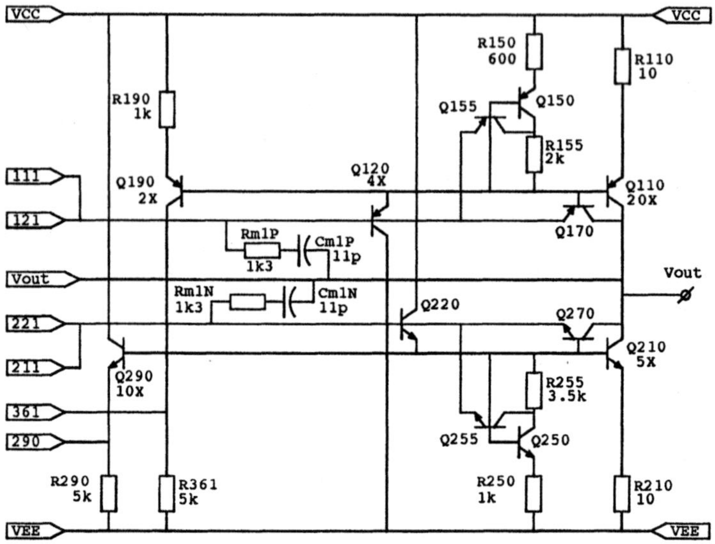

Fig. 6.13. Darlington output stage of the OpAmp on the U2010 chip.

Fig. 6.13 shows the Darlington output stage of one of the four OpAmps. The p-n-p output transistor Q_{110} and n-p-n output transistor Q_{210} are biased at a quiescent current of 320 μA. The emitter resistors R_{110} and R_{210}, have been inserted to prevent breakthrough of the output transistors at high current and voltage levels, but they do not seriously limit the output-voltage range in normal operation. The drivers Q_{120} and Q_{220}, which precede the output transistors, boost the current gain and enable the output stage to push and pull output currents well over 10 mA. The current flowing through transistors Q_{190} and Q_{290}, which are part of the class-AB current control loop, is 20 μA. As before, saturation protectors Q_{170} and Q_{270} have been inserted to protect the output transistors against heavy saturation. The output current is sensed by the current limiters Q_{150}, Q_{155} and Q_{250}, Q_{255}, which were introduced in Section 4.4. They limit the output at +/-18 mA.

The total current dissipated by the output stage of Fig. 6.13 is 395 μA, and

Miller capacitors C_{m1P} and C_{m1N} give it a bandwidth of 7.5 MHz. The resistors in series with the Miller capacitors introduce a zero at a frequency somewhat higher than the unity-gain frequency, and this zero enhances the phase margin of the output stage. The minimal supply voltage of this output stage is 1.8 V.

6.3.2 Widlar Output Stage

The Widlar output stage is shown in Fig. 6.14. The p-n-p output transistor Q_{110} is driven by the folded Darlington transistors Q_{120} and Q_{130}, and boost circuit Q_{140}, Q_{145} and Q_{147} supplies Q_{110} with sufficient base current at high output currents. Similarly, the n-p-n output transistor Q_{210} is preceded by Q_{220} and Q_{230}, and the boost circuit here consists of Q_{240}, Q_{245} and Q_{247}. The class-AB current control transistors Q_{190} and Q_{290} and the resistors R_{190}, R_{361} and R_{290} are identical to those of the Darlington output stage, and since the output transistors Q_{110} and Q_{210} are also equally as large as those of the Darlington output stage, the quiescent current through the output transistors is the same for both output stages, viz. 320 μA. Current source Q_{797}, with current mirror Q_{855}, Q_{860} ensures a minimum current of 20 μA through driver Q_{120} of the p-n-p half of the output stage, if the other, n-p-n half is driven hard. On the other hand, Q_{220} is kept also at a minimum of 20 μA by current source Q_{798}, if the p-n-p half is driven hard.

The saturation protection of Q_{110} and Q_{210} again consists of Q_{170} and Q_{270} respectively, and Q_{150}, Q_{155} and Q_{250}, Q_{255} limit the output current supplied by the p-n-p half and the n-p-n half, respectively, to +/-20 mA.

To diminish complex poles in the frequency response of the Widlar output stage, additional Miller capacitors C_{m0P} and C_{m0N} of 5 pF each have been inserted. The complete output stage is compensated with C_{m1P} and C_{m1N}, which are 13 pF, slightly larger than the corresponding capacitors in the Darlington output stage. The bandwidth of the output stage is therefore fixed at a somewhat lower value of 6.7 MHz.

The total current dissipation of the output stage is 695 μA, which, apart from the 300 μA through the output transistors, is imputable to the large currents through the folded Darlington drivers Q_{120}, Q_{130} and Q_{220} and Q_{230}. The current

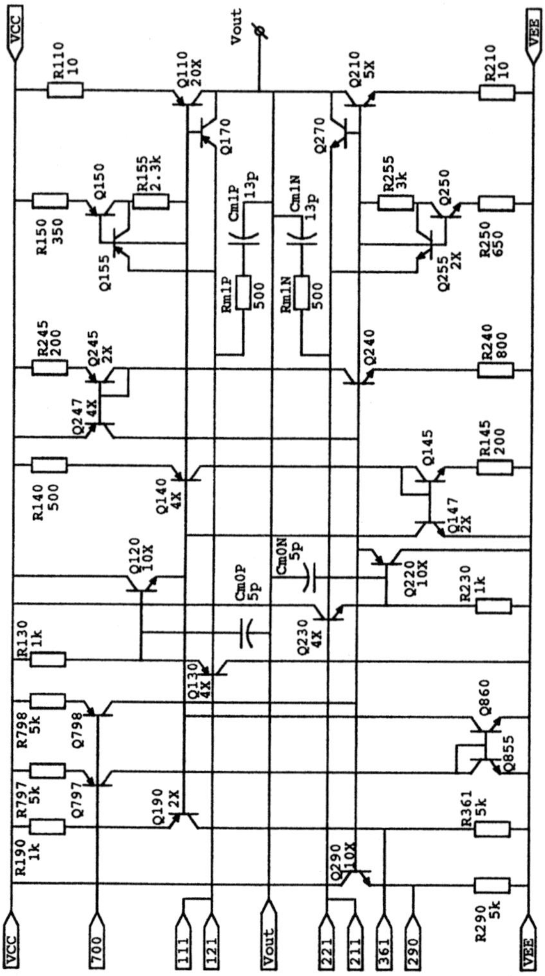

Fig. 6.14. Widlar output stage of the OpAmp on the U2011 chip.

through these transistors had to be chosen large, in order to minimize the bandwidth reduction they cause. The Widlar output stage has a minimal supply voltage of 1 V.

6.3.3 Multi-Path-Driven Output Stage

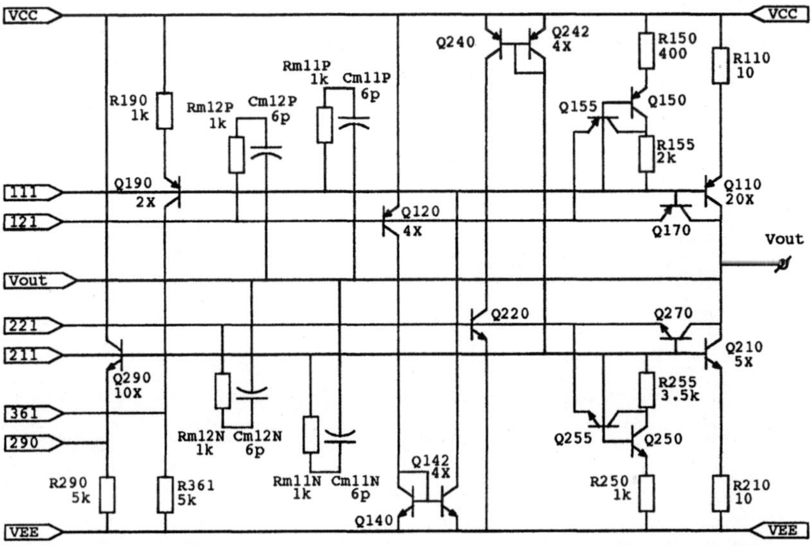

Fig. 6.15. Multi-path-driven output stage of the OpAmp on the U2010 chip.

Fig. 6.15 depicts the multi-path-driven output stage. This type of output stage was introduced in Section 3.4. The output transistors Q_{110} and Q_{210} are driven by Q_{120}, Q_{140}, Q_{142} and Q_{220}, Q_{240}, Q_{242}, respectively, which gives the output stage a sufficiently high gain and enables it to drive large output currents. To further boost the gain, the current mirrors Q_{140}, Q_{142} and Q_{240}, Q_{242} have been scaled 1:4. Parallel to this first path, Q_{110} and Q_{210} are also directly accessed by the intermediate stage, thereby by-passing the frequency-response deterioration caused by Q_{120}, Q_{140}, Q_{142} and Q_{220}, Q_{240}, Q_{242}. The class-AB current control transistors Q_{190} and Q_{290} again fix the quiescent current through the transistors at 320 μA.

Transistors Q_{170} and Q_{270} protect Q_{110} and Q_{210}, respectively, against heavy saturation, and the current-limiting transistors Q_{150}, Q_{155} and Q_{250}, Q_{255} limit the output current at +/-18 mA.

The frequency compensation of the output stage is realized with the four Miller capacitors C_{m11P} and C_{m12P}, which are connected between the output and the two inputs of the p-n-p half of the circuit, and C_{m11N} and C_{m12N}, which are connected between the output and the two inputs of the n-p-n half. The total value of these capacitors is 24 pF. The resulting unity-gain frequency of the intermediate and output stage is 7.5 MHz. The total current consumption of the multi-path-driven output stage, if no output current is supplied, is 420 μA, and the minimal supply voltage is 1 V.

6.3.4 Parallel-Feedback-Current Compensated Output Stage

Fig. 6.16 shows the output stage with parallel-feedback current compensation, the last of the four output stages. The operation of this output stage was described in Section 3.5. A second p-n-p output transistor, Q_{120} is placed parallel to Q_{110}. The emitter area of Q_{110} is about four times that of Q_{120}. The base current of Q_{120} is supplied by the feedback amplifier Q_{130}-Q_{140}, and this amplifier also steers a proportionally larger base current to Q_{110}. This is realized by scaling the emitters of the transistors that directly drive Q_{110} and Q_{120}, viz. Q_{138} and Q_{140}, respectively, with the same ratio as Q_{110} and Q_{120}. The tail current of the feedback amplifier Q_{130}-Q_{140} is supplied by the Q_{142}-Q_{148}. This current is directly proportional, but ten times smaller than the current through Q_{110}. The tail-current source could not be connected directly to the base of Q_{110}, because that would lead to saturation of Q_{148}; after all, the bases of Q_{130} and Q_{132} are also connected to the base of Q_{110}. The n-p-n part of the output stage similarly consists of the output transistors Q_{210} and Q_{220}, and of feedback amplifier Q_{230}-Q_{240}, which in turn is gets its tail current from Q_{242}-Q_{248}. This tail current is as small as 1/25-th of the current through Q_{210}. This is necessary to lower the transconductance of the feedback amplifier Q_{230}-Q_{240}, and thus ensure its stability. As explained in Section 3.5, both the feedback amplifiers Q_{130}-Q_{140} and Q_{230}-Q_{240} also need compensation

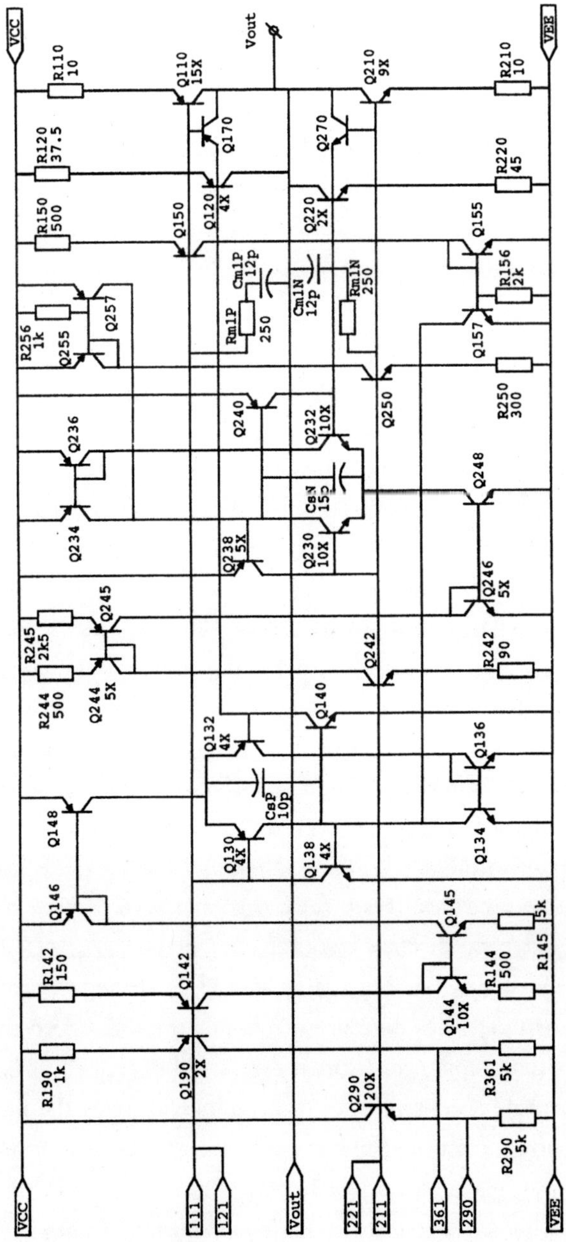

Fig. 6.16. Output stage with parallel-feedback current compensation, integrated on the U2011 chip.

capacitors C_{sP} and C_{sN}, which have values of 10 and 15 pF, respectively. The quiescent current through the output transistors Q_{110}, Q_{120} and Q_{210}, Q_{220} is again regulated by class-AB current-control transistors Q_{190} and Q_{290} at 320 μA.

The saturation-protection transistors Q_{170} and Q_{270} have been inserted in the same way as before. The current limiter is implemented differently, and now takes away the base current of Q_{138}, Q_{140} and Q_{238}, Q_{240}, to limit excessive currents through the p-n-p and n-p-n output half, respectively, at a value of +/-18 mA.

The Miller capacitors C_{m1P} and C_{m1N} compensate the output stage, and give it a bandwidth of 6 MHz. The total current dissipated by the output stage is 468 μA, and the minimum supply voltage is 1 V.

6.3.5 Common Circuit Parts

The four output stages that have been described above are preceded by a class-AB current control, an intermediate and input stage and a PTAT-current source that is equal for all four OpAmps. Fig. 6.17 shows the intermediate stage and class-AB current control. These circuit parts have four outputs, labelled "111", "121", "221" and "211", that are connected to the concerning output stage. As can be seen from Fig. 6.15, the multi-path-driven output stage needs these four inputs but the other output stages have only one input for each complementary half. To be able to use the same intermediate stage and class-AB current control for all four OpAmps, the four input signals to each of the output stages are reduced to two, at the input of the Darlington, Widlar and parallel-feedforward-compensated output stages, as can be seen from Figs. 6.13, 6.14 and 6.16.

The intermediate stage consists of two separate parts, Q_{410}, Q_{412} and Q_{420} driving the p-n-p output stage, and Q_{430}, Q_{432} and Q_{440} driving the n-p-n output stage. Each of these two parts has current mirrors that are referenced to the same supply-voltage rail as the output transistors it drives. As explained in Section 4.1, this is done to compensate the Early effect on the transistors of the intermediate stage and class-AB current control. The current of each of the tail-current sources Q_{820} and Q_{830} is 24 μA, giving a current of 8 μA through each of the branches of both differential triples. To ensure a sufficient current through the folded

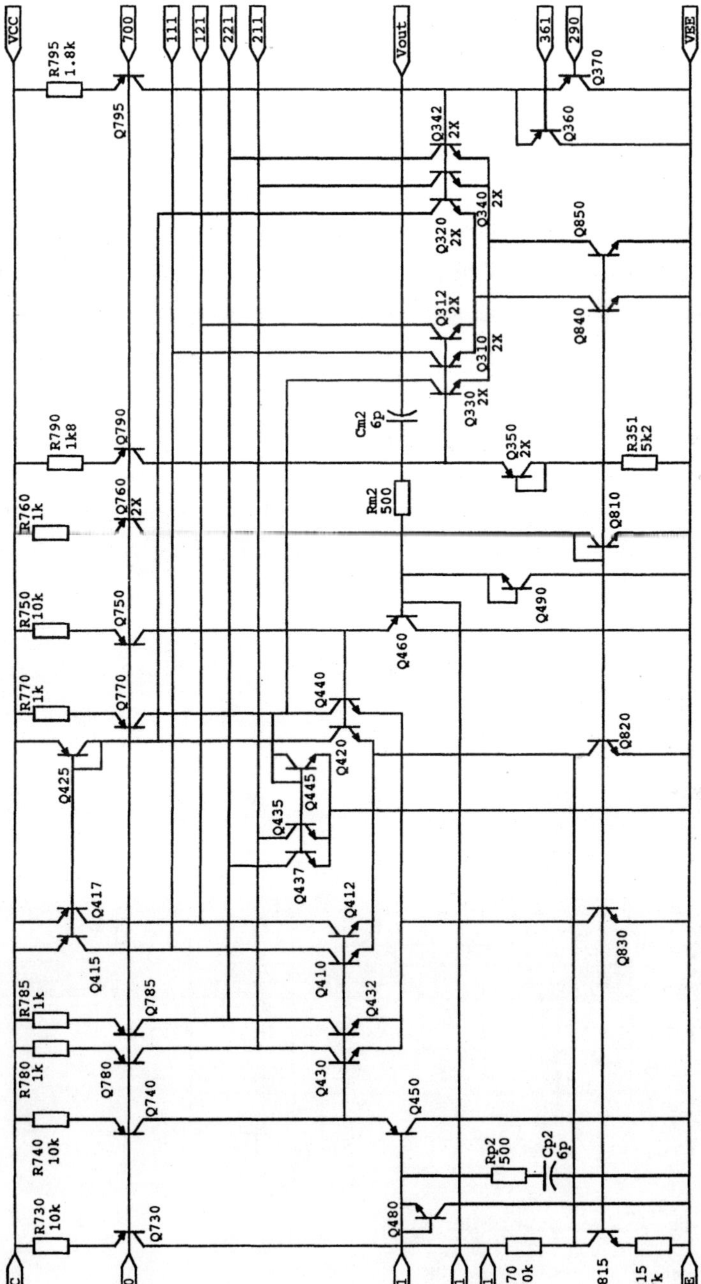

Fig. 6.17. Intermediate stage and class-AB current control of the OpAmps on the U2010 and U2011 chips.

current mirrors Q_{435}, Q_{437} and Q_{445}, also if the intermediate stage is driven hard, the current sources Q_{770}, Q_{780} and Q_{785} have to be 24 μA, equal to the tail current. The intermediate stage is cascaded with emitter followers Q_{450} and Q_{460}, to boost the current gain. The common-mode voltage on the bases of Q_{450} and Q_{460} is controlled by the feedback loop comprising the intermediate-stage transistors, R_{470} and the transistors of the summing circuit of the input stage. The voltage over resistor R_{470} is 420 mV, fixing the collector voltage of the current sources Q_{820} and Q_{830} at 270 mV above V_{EE}. The diodes Q_{480} and Q_{490} clamp the base voltages of Q_{450} and Q_{460}, respectively, to prevent saturation of current sources Q_{740} and Q_{750}, if the inputs of the OpAmp are driven hard.

The decision pair Q_{360}, Q_{370} to the right of Fig. 6.17 compares the currents flowing through the output transistors, by measuring the voltages over R_{361} and R_{290} in the output stage. The smallest of the currents through output transistors Q_{110} and Q_{210} is regulated at a quiescent current, determined by the reference voltage of feedback amplifier Q_{310}-Q_{342}. This reference voltage is generated over diode Q_{350}, matched to Q_{360} and Q_{370}, and over R_{351}. The feedback amplifier is

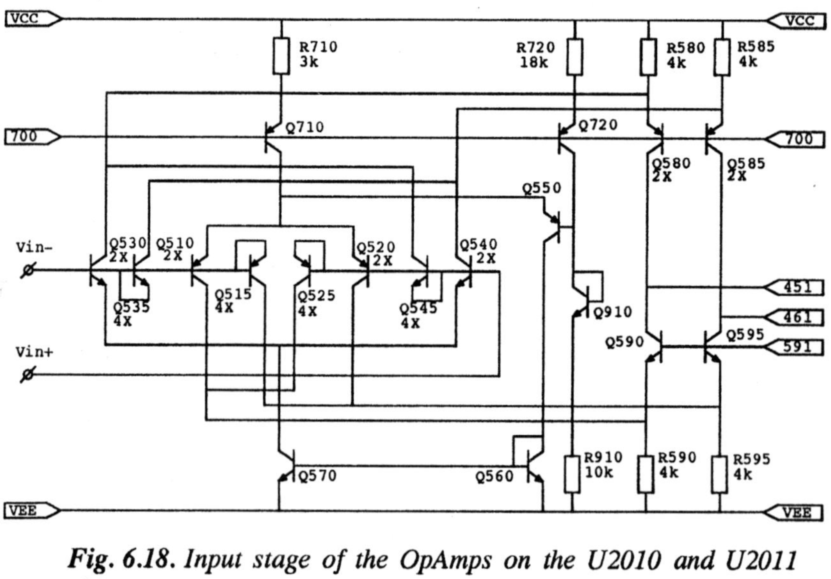

Fig. 6.18. Input stage of the OpAmps on the U2010 and U2011 chips.

composed of two parts, as is the intermediate stage, to compensate the Early effect. The differential triple Q_{310}, Q_{312}, and Q_{320} is connected to the p-n-p output stage, and the triple Q_{330}, Q_{332}, and Q_{340} is connected to the n-p-n output stage. Through each of the two parts of the feedback amplifier flows a current of 24 μA from the tail-current sources Q_{840} and Q_{850}, respectively. To prevent saturation of these current sources, the emitter area of the transistors Q_{310}-Q_{342} has been doubled.

The input stage of the four OpAmps, shown in Fig. 6.18, is a complementary, rail-to-rail input stage with constant transconductance over the complete supply-voltage range. It is very similar to the input stage of the NE5234, described in Section 6.1. Depending on the common-mode input voltage, current switch Q_{550} directs the 13-μA tail current of Q_{710} either to the p-n-p or to the n-p-n input pair. The p-n-p input pair Q_{510}, Q_{520} handles input signals in the lower common-mode range, and the n-p-n input pair Q_{530}, Q_{540} the ones in the upper common-mode range. At supply voltages below 1.8 V, only the p-n-p input pair is able to operate in the common-mode range near the negative supply rail. The collector currents of both input pairs are summed by Q_{580}-Q_{595}. The current through the branches of this summing circuit is 7.5 μA, if the input stage is in equilibrium. The diodes Q_{515}, Q_{525}, Q_{535} and Q_{545} prevent reversal of the output signal, if the common-mode input voltage largely exceeds the supply-voltage rails.

The reference-current source of the OpAmps is shown in Fig. 6.19. It comprises the PTAT-current source Q_{10}-Q_{60} and start-up circuit Q_{70}, Q_{80}, and it is similar to that of the IS549, described in the Section 6.2. The PTAT current is 6 μA, and the circuit is compensated with a 5-pF C_{c1}.

The 250 μmho input-stage transconductance, and the capacitors C_{m2} and C_{p2} of 6 pF each, give the OpAmps with Darlington and multi-path-driven output stages a unity-gain bandwidth of 3.4 MHz. The Widlar output stage has a bandwidth that is slightly lower, so the value of C_{m2} and C_{p2} is chosen 7 instead of 6 pF. The resulting unity-gain bandwidth of the OpAmp with Widlar output stage is 2.7 MHz. The OpAmp with parallel-feedback current compensation has a yet lower bandwidth, and therefore C_{m2} and C_{p2} are now 8.5 pF, yielding a bandwidth of 2.6 MHz.

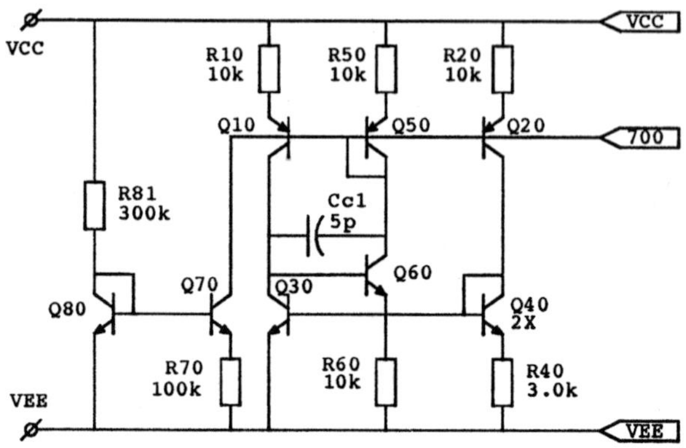

Fig. 6.19. PTAT-current generator of the OpAmps on the U2010 and U2011 chips.

6.3.6 Measurement Results

To be able to compare the performance of the four output stages, simply measuring their bandwidth is not enough. After all, the design of the output stages emanates from equal currents through their output transistors, and therefore the bandwidths of each of the four OpAmps do not differ very much. The amount of current needed in the remaining part of the output stages, apart from the output transistors, does differ, however. Therefore, besides the bandwidth, also the dissipated current in the output stage should be taken into consideration. Since we strive for the lowest supply voltage, this is also of importance. A Figure of merit is thus defined as the ratio of the bandwidth of the OpAmp and the power it consumes, that is, the product of the minimum supply voltage and the supply current. This is a slightly different definition from that given in [3], where the bandwidth and power consumption of only the output stage is considered. In formula the bandwidth-to-power ratio of the OpAmp is:

$$F_M = \frac{B_W}{V_{CC_{min}} \cdot I_{CC}} \tag{6.1}$$

This unit of this Figure of merit, Hz/W, is the inverse of the well-known power-delay product in digital circuits.

Table 6.3 gives the values of the Figure of merit for the four OpAmps, as well as the parameters from which it is derived. In accordance with the findings set out in Chapter 3, the performance of the OpAmp with multi-path-driven output stage appears better than that of the other OpAmps.

The measured open-loop frequency responses of the four OpAmps are shown in Figs. 6.20, 6.21, 6.21 and 6.22. The bandwidth of the OpAmp with Darlington output stage is 3.5 MHz, with a phase margin of 63°, and that of the OpAmp with Widlar output stage is 2.7 MHz, with a phase margin of 63°. The OpAmp with the multi-path-driven output stage has a bandwidth of 3.4 MHz, and 61° phase margin, and finally, the bandwidth of the OpAmp with parallel-feedback-current-compensated (PFCC) output stage is 2.5 MHz, with 61° phase margin.

The specifications of the four OpAmps are summarized in Table 6.4. Most specs were discussed earlier, and all are in accordance with the expectations on

Table 6.3. *Figure of merit, $F_M = B_w / V_{CC\,min} I_{CC}$, for the four OpAmps. The abbreviation MPD and PFCC stand for multi-path-driven, and parallel-feedback-current-compensated, respectively.*

Parameter	Darlington	Widlar	MPD	PFCC	Unit
BPR (*B*andwidth *P*ower-*R*atio)	2.9	2.8	4.9	3.5	MHz/mW
Bandwidth	3.5	2.7	3.4	2.6	MHz
Minimum Supply Voltage	1.8	1.0	1.0	1.0	V
Supply Current	670	980	700	750	μA

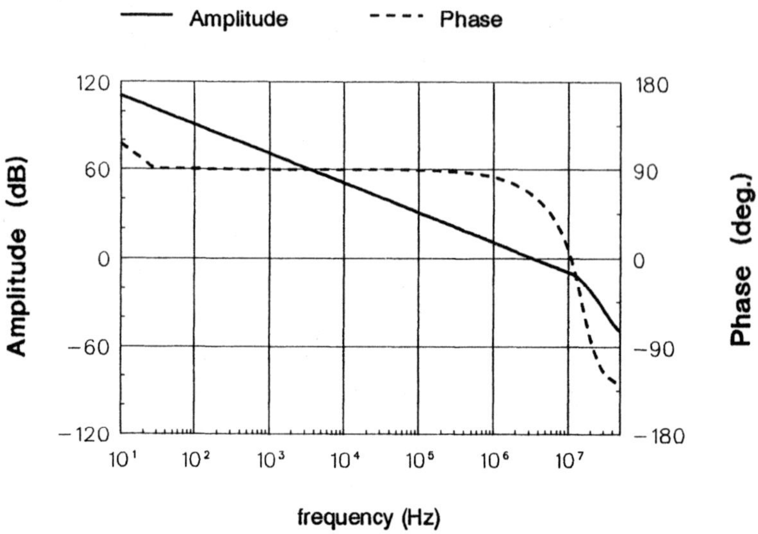

Fig. 6.20. Open-loop frequency response of the OpAmp with Darlington output stage. V_{sup}= 2 V, T= 27 °C, R_L= 10 kΩ, C_L= 100 pF.

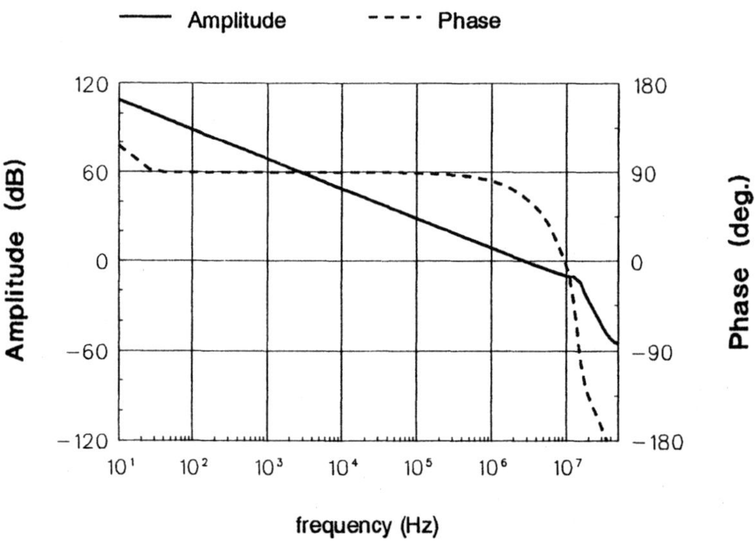

Fig. 6.21. Open-loop frequency response of the OpAmp with Widlar output stage. V_{sup}= 1 V, T= 27 °C, R_L= 10 kΩ, C_L= 100 pF.

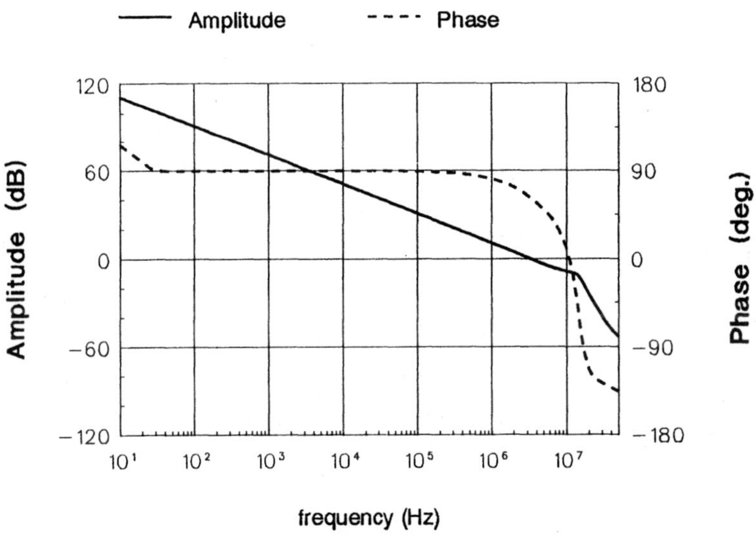

Fig. 6.22. Open-loop frequency response of the OpAmp with multi-path-driven output stage. V_{sup}= 1 V, T= 27 °C, R_L= 10 kΩ, C_L= 100 pF.

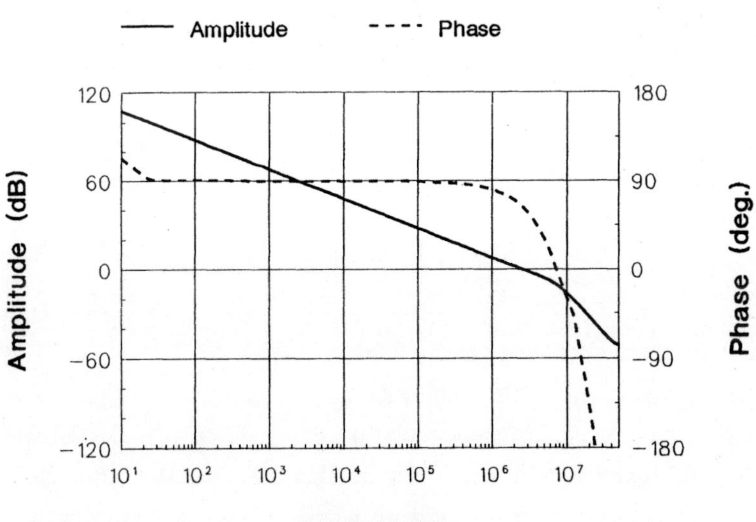

Fig. 6.23. Open-loop frequency response of the OpAmp with parallel-current-compensated output stage. V_{sup}= 1 V, T= 27 °C, R_L= 10 kΩ, C_L= 100 pF.

Table 6.4. Specifications of the four OpAmps. V_{sup} *= 1 V, (2 V for Darlington) T= 27 °C,* R_L *= 10 kΩ,* C_L *= 100 pF.*

symbol	Darlington	Widlar	MPD	PFCC	unit
T_{OP}	-55/125	-55/125	-55/125	-55/125	°C
V_{CC}	1.8-10	1-10	1-10	1-10	V
I_{CC}	670	980	700	750	μA
V_{OS}	0.6	0.6	0.6	0.6	mV
I_B	140	140	140	140	nA
V_{CM}	V_{EE}-0.25/ V_{CC}+0.25	V_{EE}-0.25/ V_{CC}+0.25	V_{EE}-0.25/ V_{CC}+0.25	V_{EE}-0.25/ V_{CC}+0.25	V
CMRR	100	100	100	100	dB
I_L	±15	±15	±15	±15	mA
A_{VOL}	110	115	117	115	dB
V_{OUT}	V_{EE}+0.1/ V_{CC}-0.1	V_{EE}+0.1/ V_{CC}-0.1	V_{EE}+0.1/ V_{CC}-0.1	V_{EE}+0.1/ V_{CC}-0.1	V
S_R	1.1	0.9	1.1	0.85	V/μs
BW	3.5	2.6	3.4	2.6	MHz
θ_M	63	63	61	61	°
NOISE	23	23	23	23	nV/√Hz

account of the theory discussed in the previous Chapters.

The micrograph of the U2010 chip is shown in Fig. 6.24. The OpAmp with the Darlington output stage is placed on top of the chip, and below it the OpAmp with the multi-path-driven output stage can be seen. The U2011 chip, shown in Fig. 6.25, contains the OpAmp with parallel-feedback-current compensation (top) and the OpAmp with Widlar output stage (bottom). On the right-hand side of the chips, the output stages can be distinguished, in the middle the intermediates and class-AB current controls are situated. The input stages with quad layout can be

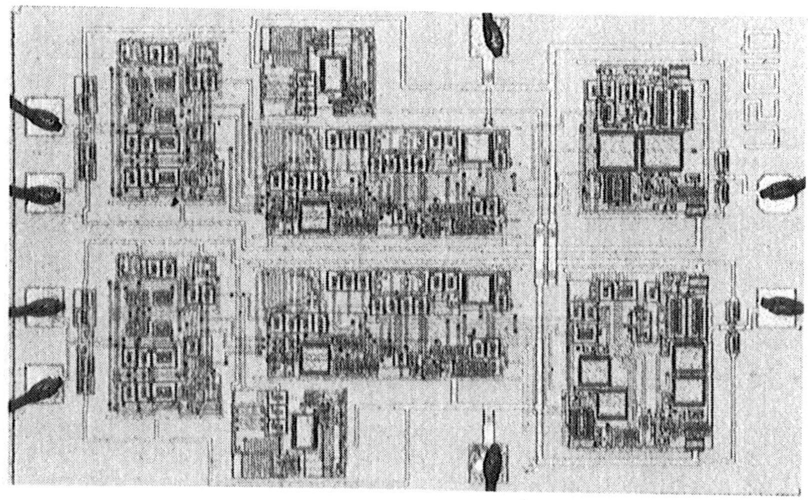

Fig. 6.24. Micrograph of the U2010 chip. On top the OpAmp with Darlington output stage, below the OpAmp with multi-path-driven output stage.

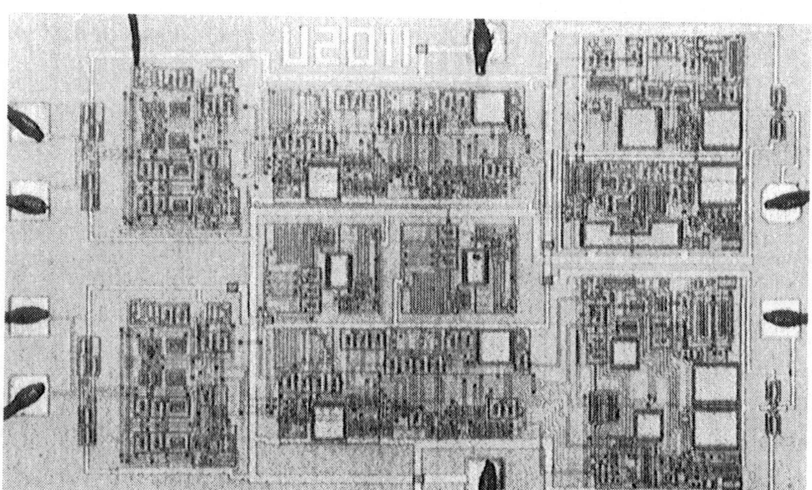

Fig. 6.25. Micrograph of the U2011 chip. On top the OpAmp with parallel-feedback-current compensated output stage, below the OpAmp with Widlar output stage.

seen to the left of the micrographs. Each of the two chips measures 2.2 × 1.4 mm^2 (87 × 55 mil^2), of which 0.20 mm^2 (310 mil^2) is occupied by the Darlington output stage, 0.26 mm^2 (400 mil^2) by the multi-path-driven output stage, 0.35 mm^2 (540 mil^2) by the parallel-feedback-current compensated output stage and 0.30 mm^2 (470 mil^2) by the Widlar output stage.

6.4 1-V OPAMP WITH 10-MHz BANDWIDTH

The comparison of the four output stages in the previous Section showed that at a 1-V supply voltage, the multi-path-driven output stage gives the best performance. A second OpAmp has therefore been designed comprising this output stage, but now in the QUBiC process of the Signetics Company, Sunnyvale CA. The QUBiC process is a 7-V BICMOS process, with oxide-isolated transistors. The n-p-n transistors have a transit frequency of 13 GHz and the f_T of the p-n-p's is 200 MHz. The OpAmp design presented in this Section is intended as a system-library building block. Because of this, the maximum output current of the OpAmp could be chosen smaller than in the designs presented in the preceding Sections, and the maximum load capacitor is also smaller. However, the bandwidth of the circuit should be as high as 10 MHz, and the total current consumption should be lower.

Fig. 6.26 shows the output stage of the 10-MHz OpAmp. The p-n-p output transistor Q_{110} is, through the bus labelled "111", directly connected to the intermediate stage. To boost the gain of the output stage, there is also a second signal path through Q_{120}, Q_{140} and Q_{142}. This path is connected to the preceding stages by the bus labelled "121". Similarly, the n-p-n output half consists of output transistor Q_{210}, which is driven by Q_{220}, Q_{240} and Q_{242}. Transistor Q_{220} is connected to the intermediate stage by the bus "221", and parallel to this path the feedforward path "211" directly connects the intermediate stage to the n-p-n output transistor. As can be seen from Fig. 6.26, the topology of the output stage is identical to that of the multi-path-driven output stage on the U2010 chip, discussed

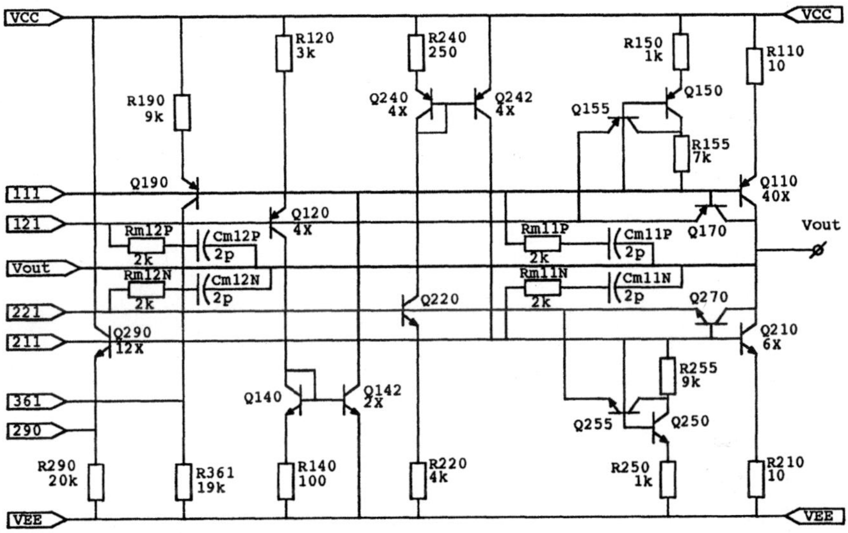

Fig. 6.26. Multi-path-driven output stage of the 10-MHz OpAmp.

in Section 6.3.3. The p-n-p's in that circuit are vertical transistors with an f_T of 1 GHz, whereas the p-n-p's in the QUBiC process are 200-MHz lateral transistors. According to the theory presented in Section 3.4, this nevertheless yields a high-frequency output stage, as long as the transconductance of Q_{120}-Q_{144} and Q_{220}-Q_{242} is kept low. To this end, resistors R_{120} and R_{220} have been inserted in series with Q_{120} and Q_{220}, respectively. Transistors Q_{190} and Q_{290} are part of the class-AB current-control loop that defines a quiescent current of 100 μA through the output transistors Q_{110} and Q_{210}. To prevent deep saturation of the output transistors, Q_{170} and Q_{270}, which sense the voltage at the output, are inserted and, if necessary, they remove the driving current of the transistors Q_{120} and Q_{220}, respectively. The output current is limited to +/-5 mA by the current limiters Q_{150}, Q_{155} and Q_{250}, Q_{255}, which were introduced in Section 4.4.

The intermediate stage and class-AB current control are depicted in Fig. 6.27. The topology of this circuit part, too, is almost identical to that of the OpAmps on the U2010 and U2011 chips. The intermediate stage consists of two triples, Q_{410}, Q_{412} and Q_{420} connected to the p-n-p output stage, and Q_{430}, Q_{432} and Q_{440} connected to the n-p-n output stage. Each half has its own current source Q_{820} and

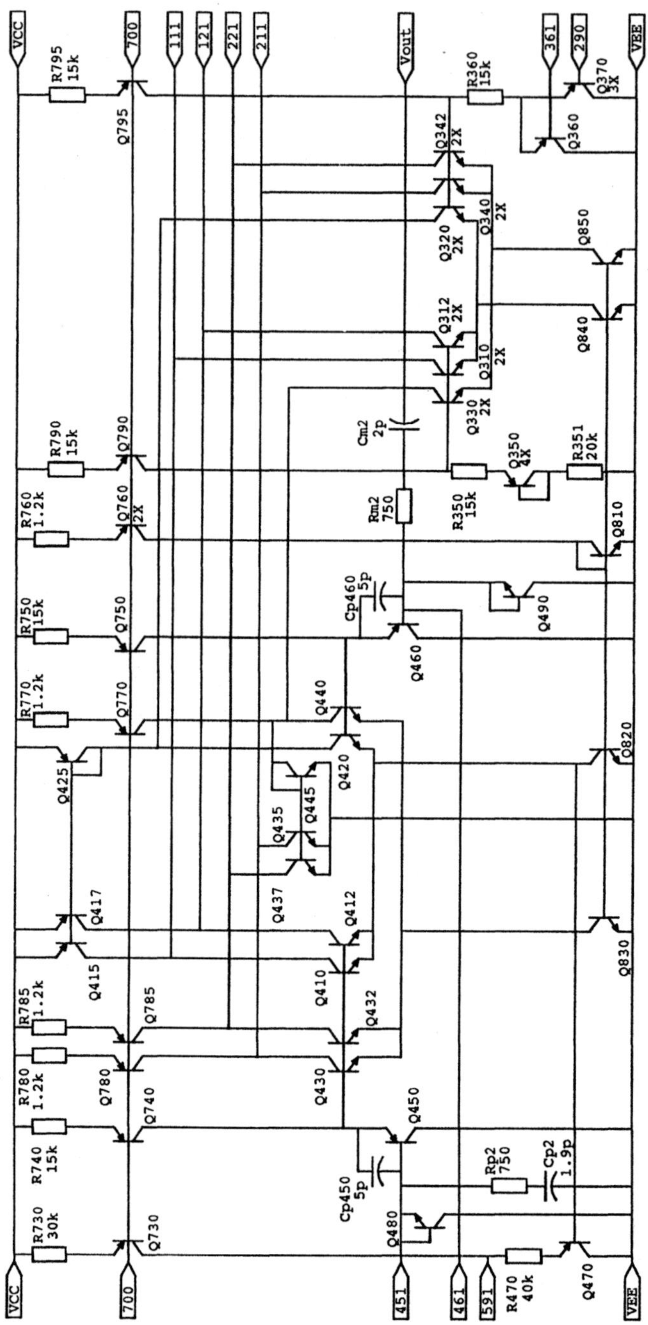

Fig. 6.27. Intermediate stage and class-AB current control of the 10-MHz OpAmp.

Q_{830} of 25 μA each, and its own current mirrors, Q_{415}, Q_{417} and Q_{425}, and Q_{435}, Q_{437} and Q_{445}. The folded emitter followers Q_{450} and Q_{460} increase the input resistance of the intermediate stage and thereby enlarge the gain of the OpAmp. To minimize the influence on the frequency behavior, the feedforward capacitors C_{p450} and C_{p460} shunt Q_{450} and Q_{460}. The common-mode input-voltage level at the input of the intermediate stage is fixed at 200 mV above V_{EE} with the biasing loop through the intermediate-stage transistors, Q_{470}, R_{470} and the transistors of the input-stage summing circuit. The voltage on the collectors of the current sources Q_{820} and Q_{830} is 100 mV above V_{EE}. The diodes Q_{480} and Q_{490} prevent saturation of current sources Q_{740} and Q_{750} if the inputs of the OpAmp are driven hard. The class-AB current-control feedback amplifier also consists of two parts, Q_{310}, Q_{312} and Q_{320} connected to the p-n-p output stage, and Q_{330}, Q_{332} and Q_{340} to the n-p-n output stage. The current sources Q_{840} and Q_{850} have a value of 25 μA each. Transistors Q_{360} and Q_{370} select the smallest of the currents through the output transistors. Feedback amplifier Q_{310}-Q_{340} then regulates this smallest of the two at a value determined by the reference voltage over diode Q_{350}. Because Q_{350} is matched to Q_{360} and Q_{370}, the quiescent current through the output transistors is insensitive to variations in temperature. The resistors R_{350} and R_{360} are inserted to shift the voltage level at the bases of the feedback amplifier, thereby giving the current sources Q_{840} and Q_{850} a little more headroom. Transistors Q_{360} and Q_{370} are scaled 1:3. This has the effect that if the p-n-p output stage is driven hard, the rest current through the n-p-n output transistor Q_{210} is relatively large when compared with the rest current through the p-n-p transistor if the n-p-n output stage is driven hard. The idea is that the n-p-n output half should still contribute substantially to the frequency response of the complete output stage, even if the p-n-p output half is supplying the DC output current.

Fig. 6.28 shows the input stage of the 10-MHz OpAmp. The collector currents of the n-p-n input pair Q_{510}, Q_{520} and the n-p-n input pair Q_{530}, Q_{540} are added by the summing circuit Q_{580}-Q_{595}. The diodes Q_{515}, Q_{525}, Q_{535} and Q_{545} prevent reversal of the output signal of the OpAmp, if the common-mode input voltage exceeds the supply rails by more than 300 mV, and current switch Q_{550} directs the 10 μA tail current from tail-current source Q_{710} either to the p-n-p input pair, or, if the common-mode input voltage is higher than the reference voltage of 0.75 V,

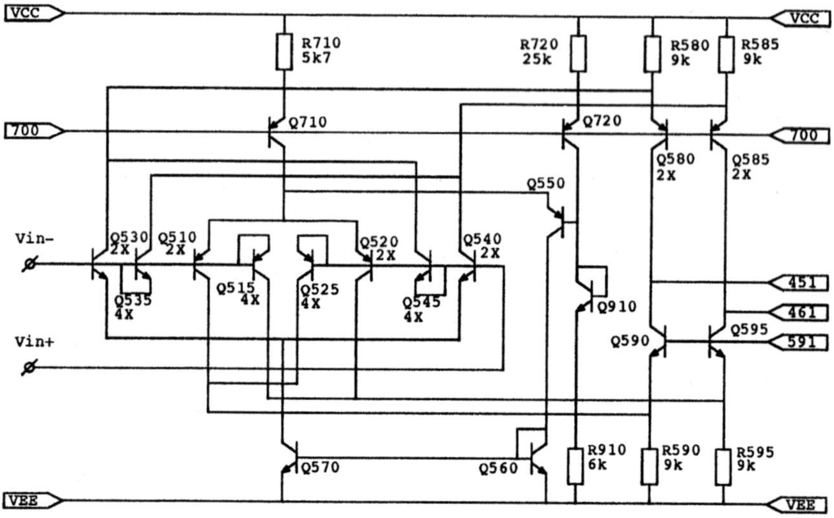

Fig. 6.28. Input stage of the 10-MHz OpAmp.

determined by diode Q_{910}, to the n-p-n input pair. This implies that at a 1 V supply voltage, only the p-n-p input pair is able to operate, in the common-mode input range near the negative supply rail. As has been explained in Chapter 2, the OpAmp has a rail-to-rail common-mode input-voltage range at supply voltages above 1.8 V.

The current through the branches Q_{10}-Q_{30} and Q_{20}-Q_{40} of the PTAT current source, shown in Fig. 6.29, is equal to 5 μA. The diode Q_{50} and transistor Q_{60} drain the base current of the current sources of the OpAmp, which is coming in from the biasing bus "700". The 0.5-μA start current from Q_{70} and Q_{80} makes sure that the PTAT current actually does flow through the circuit, and capacitor C_{c1} compensates the feedback loop around Q_{60}.

The frequency compensation of the multi-path driven output stage is realized with Miller capacitors C_{m11P} and C_{m12P}, connected between the two inputs of the p-n-p half of the circuit and the output, and C_{m11N} and C_{m12N}, which are connected between the inputs of the n-p-n half and the output. The value of the capacitors is 2 pF, and in series with them 2-kΩ resistors have been inserted to improve the phase margin of the output stage. The unity-gain bandwidth of the output stage

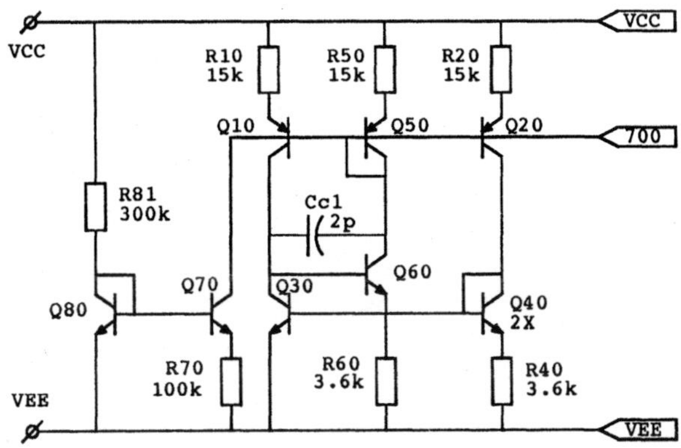

Fig. 6.29. PTAT-current generator of the 10-MHz OpAmp.

and intermediate stage is 20 MHz. The pole that is added if the input stage is placed before the output and intermediate stage, is split with C_{m2} and C_{p2}. These capacitors, of 2 and 1.9 pF, respectively, fix the overall unity-gain bandwidth of the OpAmp at 10 MHz. In series with C_{m2} and C_{p2} the resistors R_{m2} and R_{p2}, of 750 Ω each, have been inserted to increase the phase margin of the OpAmp to more than 60°.

At the time of writing, the 10-MHz OpAmp had not yet been processed. The data presented here is therefore result of extensive simulation, rather than of measurement. The open-loop frequency response is shown in Fig. 6.30. The supply voltage is 1 V and the OpAmp is loaded with 10 kΩ parallel to 10 pF. As can be seen from Fig. 6.30, the bandwidth of the OpAmp is 10 MHz and the phase margin is 64°.

Table 6.5 summarizes the expected specifications of the OpAmp. The 10-MHz OpAmp has a temperature range from -55 to +125 °C, although the supply voltage has to be increased to 1.2 V to be able to operate the OpAmp at temperatures below 0 °C. The total supply current is 390 μA, and it is constant within 15% over the supply voltage and temperature range. The input bias current depends on whether the p-n-p or the n-p-n input pair is operating; the input bias current of the p-n-p input pair is 120 nA, that of the n-p-n pair is 45 nA. As explained before,

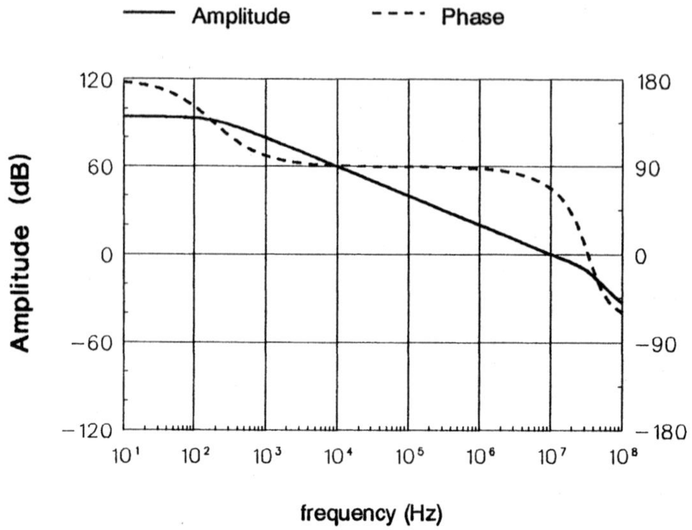

Fig. 6.30. *Simulated open-loop frequency response of the 10-MHz OpAmp.* V_{sup}*=1 V, T= 27 °C,* R_L*= 10 kΩ,* C_L *10 pF.*

the common-mode input-voltage range is limited to the range near the negative supply rail at a supply voltage of 1 V, but at supply voltages higher than 1.8 V, the input range reaches from rail to rail, and even exceeds the supply rails. The maximum load current that can be sourced or sunk is limited to +/-2 mA at a 1-V supply. After that, the transistors Q_{120} and Q_{220} start to saturate, because of the voltage drop over R_{120} and R_{220}, respectively. At higher supply voltages, this does not happen, and a larger load current can be supplied.

Table 6.5. Simulated specifications of the 10-MHz OpAmp. $V_{sup}= 1\ V$, $T= 27\ °C$, $R_L= 10\ k\Omega$, $C_L= 10\ pF$.

symbol	parameter	value	unit
T_{OP}	operation temperature range	-55/125	°C
V_{CC}	supply voltage range	1-7.5	V
I_{CC}	supply current	390	μA
V_{OS}	offset voltage	0.5	mV
I_B	input bias current	45-120	nA
V_{CM}	common-mode input range	V_{EE}-0.25/ V_{CC}+0.25	V
CMRR	common-mode rejection ratio	100	dB
I_L	peak-load current	±2	mA
A_{VOL}	open-loop voltage gain	90	dB
V_{OUT}	output-voltage swing	V_{EE}+0.1/ V_{CC}-0.05	V
S_R	slew rate	3.0	V/μs
BW	open-loop bandwidth	10	MHz
θ_M	phase margin	64	°
NOISE	input referred noise voltage at 1kHz	15	nV/√Hz

6.5 CONCLUSIONS

The OpAmps that have been discussed in this Chapter all have a rail-to-rail common-mode input-voltage range, provided that the supply voltage is higher than 1.8 V. The OpAmp IS549 has this feature also at supply voltages down to 1 V; this has been achieved at the cost of a deterioration of parameters such as the offset, the *CMRR* the noise and the occupied chip area. The output stages of all OpAmps have an output-voltage range that reaches from one supply rail to the other, and that allow connection of the output to either one of the supply rails. The output current that can be supplied by the OpAmps is larger than 10 mA, except for the 10-MHz OpAmp, which has been specifically designed for applications where smaller output currents are sufficient. The over-all gain of the OpAmps is larger than 100 dB, and the total supply current of most of the OpAmps is in the order of 700 μA. The bandwidth of the OpAmps depends on the quiescent current through the output stage and on the load capacitor that should be driven. Further, it appeared that, in accordance with the theory presented earlier, that the best frequency behavior is accomplished if the signal path to the output transistor is as

***Table 6.6**. Bandwidth-to-Power ratio of the realized OpAmps.*

OpAmp	*BPR* (MHz/mW)	B_W (MHz)	V_{CC} (V)	I_{CC} (μA)
NE5234	1.8	2.5	2	690
IS549	0.6	0.45	1	750
U2010, Darlington	2.9	3.5	1.8	670
U2010, MPD	4.9	3.4	1	700
U2011, Widlar	2.8	2.7	1	980
U2011, PFCC	3.5	2.6	1	750
10-MHz OpAmp	26	10	1	390

short as possible. At a supply voltage of 1 V, when the Darlington output stage cannot be used, the best performance is achieved by the multi-path-driven output stage. The two OpAmps with this output stage, described in Sections 6.3.3 and 6.4, have a bandwidth of 3.4 and 10 MHz, respectively.

The conclusion can then be reached that the aims set out in the Introduction have been achieved, although in none of the designs presented in this Chapter are all features combined.

The bandwidth-to-power ratio that was defined in Section 6.3.6 offers a Figure of merit to compare the performance of the OpAmps. Table 6.6 summarizes the *BPR* of the OpAmps that have been discussed in this Chapter. The *BPR* of the 10-MHz OpAmp is by far the highest, but then again this OpAmp can be loaded with a capacitor of 10 pF, maximum, whereas the load is 100 pF for the others.

6.6 REFERENCES

[1] J. H. Huijsing and D. Linebarger, "Low-Voltage Operational Amplifier with Rail-to-Rail Input and Output Ranges", *IEEE J. Solid-State Circuits*, Vol. SC-20, pp. 1144-1150, Dec. 1985.

[2] J. Fonderie, M.M. Maris, E. J. Schnitger and J. H. Huijsing, "1-V Operational Amplifier with Rail-to-Rail Input and Output Ranges", *IEEE J. Solid-State Circuits,* Vol. SC-24, pp. 1551-1559, Dec. 1989.

[3] J. Fonderie, J. H. Huijsing, "Operational Amplifier with 1-V Rail-to-Rail Multi-path-Driven Output Stage", *IEEE J. Solid-State Circuits*, Vol. SC-26, pp. 1817-1824, Dec. 1991.

CHAPTER 7
CONCLUSIONS

In this last Chapter, the objectives that were set out in the Introduction are verified, and the major conclusions of the previous Chapters are recapitulated. Finally, some references to research in adjacent areas are given.

- a rail-to-rail input-voltage range:

The low-voltage OpAmp design has to include a rail-to-rail common-mode input-voltage range. In Chapter 2 it was shown that, at supply voltages down to 1.8 V, this can be realized with a simple input stage, that in addition features a constant transconductance over the complete common-mode input range. At lower supply voltages, this input stage still is able to operate, but only in a common-mode range near the negative supply rail. If the rail-to-rail capability is also required at supply voltages down to 1 V, an input stage with a level-shift circuit can be used. If this level shift is activated, the extension of the rail-to-rail operation is exchanged for a deterioration of parameters such as the input noise voltage and offset. Furthermore, the level-shift circuit involves an increase in the chip area used. A realization of both input stages was discussed in Chapter 6.

- a rail-to-rail output-voltage range:

All output stages discussed in Chapter 3 have an output transistor in common-emitter configuration, and, consequently, the output voltage is able to reach both supply rails within about 100 mV. Moreover, the output stages are provided with protection circuits that enable it to operate with the output connected to one of the supply rails.

- an overall voltage gain of more than 100 dB:

To achieve a high overall gain of the OpAmp, an intermediate stage has to be placed between the input and output stage. This gain is then the product of the input transconductance, the intermediate-stage current gain, the output-stage current gain and the load resistance. In Chapters 2, 3 and 4 it was shown that practical values for these parameters are in the order of 0.1 mmho, 10^3, 10^3 and 10 kΩ, respectively, resulting in a total gain of 120 dB. This is well over the intended 100 dB. All the realizations described in Chapter 6 confirm this.

- *an output current drive in the order of +/-10 mA:*

In Chapter 3, it was shown, that in order to be able to supply an output current in the order of 10 mA, the current gain of the output stage has to be about 10^3. The realized OpAmps all meet this condition, and, indeed, are able to sink or source more than 10 mA. The maximum output current of the 10-MHz OpAmp, described in Section 6.4, is smaller, but this is in accordance with the intended application of this device; if necessary, the design could be altered to achieve a higher maximum output current.

- *a bandwidth in the Mega-hertz range:*

As shown in Chapters 3 and 5, the open-loop bandwidth of an OpAmp is determined by the quiescent current through the output transistors, and by the load capacitor that has to be driven. The limited transit frequency of the transistors in the signal path preceding the output transistors, especially if these are lateral p-n-p's, easily causes complex poles to occur in the frequency response of the OpAmp. To guarantee the stability of the OpAmp, the bandwidth then has to be lowered considerably, as shown in Chapter 5. It therefore appeared to be important to employ an output stage with a direct path to the output transistor. The multi-path-driven output stage, introduced in Chapter 3, meets this demand. At the expense of increased current consumption, OpAmps with other output stages can also be made to have a large bandwidth, and the relation between the bandwidth, and the power consumption of the OpAmp has been established in the Bandwidth-to-Power Ratio (*BPR*) as defined in Chapter 6. All but one of the realized OpAmps have a bandwidth in the Mega-Hertz range. The IS549 OpAmp has a bandwidth of only 450 kHz, but then again it is realized in a process with moderate 3-MHz lateral p-n-p's.

- *a total current consumption of less than 1 mA:*
The total current dissipation of the OpAmp is largely determined by the current consumption of the output stage, and this in turn is directly related to the intended bandwidth and load capacitor that has to be driven. The realized OpAmps, with bandwidths in the Mega-Hertz range, have a current consumption of about 700 μA, and thus remain below the proposed value of 1 mA.

We can, therefore, conclude that the aims stated in the Introduction have been met. There are many areas however, that are not covered in this book. These include:

- *simplified designs:*
Especially if the OpAmp is used as an on-chip ASIC building block, it is essential that it be small and economical. This requires a simplification of the OpAmp design. An OpAmp with a simplified implementation of the class-AB current control can be found in [1]. A simple design with a totally different topology, incorporating a combined intermediate stage and class-AB control is described in [2] and in [3] the same combination is used in a 1-V comparator.

- *new compensation techniques:*
The multi-path-driven compensation technique described in Chapter 5 has to date not yet been applied to a low-voltage OpAmp. This compensation method, as well as other completely different methods to compensate the OpAmp, appear to be an interesting research subject. In [4] a high-frequency and very linear OpAmp, although not operating at low supply voltages, is described.

- *CMOS and BICMOS designs:*
All the OpAmp designs in this thesis were realized in bipolar processes. New potentials and limitations arise, however, if the low-voltage design is implemented in a CMOS or BICMOS process. An approach similar to that presented here, but realized in a CMOS process, is described in [5].

REFERENCES

[1] W. C. M. Renirie and J. H. Huijsing, "Simplified Class-AB Control Circuits for Rail-to-Rail Output Stages of Operational Amplifiers", in *Proc. ESSCIRC*, 1992, pp. 183-186.

[2] M. Bredius and J. H. Huijsing, "Simplified Low-Voltage and Low-Power OpAmp with Non-Linear Class-AB Rail-to-Rail Output Stage", in *Proc. ESSCIRC*, 1992, pp. 195-198.

[3] C. J. Grootveld, J. Fonderie and J. H. Huijsing, "A 1-V Simplified Comparator with Non-Linear Class-AB Driver and Rail-to-Rail Output Stage", in *Proc. IEEE Int. Symp. on Circuits and Systems*, 1992, pp. 1380-1383.

[4] R. G. H. Eschauzier, L. P. T. Kerklaan and J. H. Huijsing, "A 100MHz 100dB Operational Amplifier with Multi-path Nested Miller Compensation Structure", in *ISSCC Dig. Tech. Papers*, 1992, pp. 196-197.

[5] R. Hogervorst, R. J. Wiegerink, P. A. L. de Jong, J. Fonderie, R. F. Wassenaar and J. H. Huijsing, "CMOS Low-Voltage Operational Amplifiers with Constant g_m Rail-to-Rail Input Stage", in *Proceedings IEEE Int. Symp. on Circuits and Systems*, 1992, pp. 2876-2879.

INDEX